Annette Heinrich

Molecular mechanisms of the effect of the mood stabilizer lithium

Annette Heinrich

Molecular mechanisms of the effect of the mood stabilizer lithium

On the mechanism of the effect of lithium on cAMP-induced CREB transcriptional activity

Südwestdeutscher Verlag für Hochschulschriften

Impressum / Imprint
Bibliografische Information der Deutschen Nationalbibliothek: Die Deutsche Nationalbibliothek verzeichnet diese Publikation in der Deutschen Nationalbibliografie; detaillierte bibliografische Daten sind im Internet über http://dnb.d-nb.de abrufbar.
Alle in diesem Buch genannten Marken und Produktnamen unterliegen warenzeichen-, marken- oder patentrechtlichem Schutz bzw. sind Warenzeichen oder eingetragene Warenzeichen der jeweiligen Inhaber. Die Wiedergabe von Marken, Produktnamen, Gebrauchsnamen, Handelsnamen, Warenbezeichnungen u.s.w. in diesem Werk berechtigt auch ohne besondere Kennzeichnung nicht zu der Annahme, dass solche Namen im Sinne der Warenzeichen- und Markenschutzgesetzgebung als frei zu betrachten wären und daher von jedermann benutzt werden dürften.

Bibliographic information published by the Deutsche Nationalbibliothek: The Deutsche Nationalbibliothek lists this publication in the Deutsche Nationalbibliografie; detailed bibliographic data are available in the Internet at http://dnb.d-nb.de.
Any brand names and product names mentioned in this book are subject to trademark, brand or patent protection and are trademarks or registered trademarks of their respective holders. The use of brand names, product names, common names, trade names, product descriptions etc. even without a particular marking in this work is in no way to be construed to mean that such names may be regarded as unrestricted in respect of trademark and brand protection legislation and could thus be used by anyone.

Coverbild / Cover image: www.ingimage.com

Verlag / Publisher:
Südwestdeutscher Verlag für Hochschulschriften
ist ein Imprint der / is a trademark of
OmniScriptum GmbH & Co. KG
Heinrich-Böcking-Str. 6-8, 66121 Saarbrücken, Deutschland / Germany
Email: info@svh-verlag.de

Herstellung: siehe letzte Seite /
Printed at: see last page
ISBN: 978-3-8381-1176-6

Zugl. / Approved by: Göttingen, Uni, Diss., 2009

Copyright © 2009 OmniScriptum GmbH & Co. KG
Alle Rechte vorbehalten. / All rights reserved. Saarbrücken 2009

Table of contents

TABLE OF CONTENTS	1
LIST OF FIGURES	8
LIST OF TABLES	10
ABBREVIATIONS	11
INTRODUCTION	**15**
1. BIPOLAR DISORDER	15
2. PHARMACOLOGY OF LITHIUM	16
2.a Pharmacokinetic of lithium	16
2.b Pharmacodynamic of lithium	17
2.b.I Cyclic AMP	18
2.b.II Phosphatidylinositol	19
2.b.III Glycogen synthase kinase 3β	19
3. THE TRANSCRIPTION FACTOR CREB	22
3.a The structure of CREB	23
3.b Transcriptional regulation mediated by CREB	24
4. THE CREB COACTIVATOR TORC	25
4.a The structure of TORC	26
4.b Regulation of TORC	27
5. LITHIUM ENHANCES cAMP-INDUCED CREB-DIRECTED GENE TRANSCRIPTION	29
6. OBJECTIVES OF THE STUDY	29
MATERIALS AND METHODS	**30**
1. EQUIPMENT	30
1.a Apparatus	30
1.b Consumables	31
1.c Kits	32
2. CHEMICALS	33
2.a Substances	33
2.b Stock solutions and buffers	36
2.b.I. Stocks	36
2.b.II. Antibiotics and protease inhibitors	37

2.b.III. General buffers 37
3. BIOLOGICAL MATERIAL..38
   3.a Bacteria strains 38
   3.b Eurcaryotic cell lines 38
   3.c Cell culture and culture media 38
      3.c.I Buffers and solutions 38
      3.c.II. Procaryotic cultures 39
      3.c.III. Eucaryotic cultures 39
   3.d Plasmids and oligonucleotides 39
      3.d.I. Expression constructs 39
      3.d.II. Luciferase reporter gene constructs 44
      3.d.III. Oligonucleotides 45
         3.d.III.A Oligonucleotides used for PCR cloning 46
         3.d.III.B Oligonucleotides used for site-directed mutagenesis 48
         3.d.III.C Oligonucleotides used for sequencing 48
         3.d.III.D Oligonucleotides used for quantitative real-time PCR 49
         3.d.III.E Oligonucleotides to generate probes 50
   3.e Enzymes and markers 51
      3.e.I. Restriction endonucleases 51
      3.e.II. Modifying enzymes 51
      3.e.III. Markers 52
   3.f Antibodies 52
4. WORKING WITH DNA AND RNA 54
   4.a DNA gel electrophoresis 54
      4.a.I. Buffers and solutions 54
      4.a.II. Gel preparation and electrophoresis 54
   4.b DNA purification from agarose gels and solutions 55
   4.c PCR and site-directed mutagenesis 55
      4.c.I. Buffer and solutions 55
      4.c.II. Polymerase chain reaction (PCR) 55
      4.c.III. Site-directed mutagenesis using primerless PCR 56
   4.d DNA modification 57
      4.d.I. Restriction digest 57
      4.d.II. Blunting 58

- 4.d.III. Ligation of DNA 58
- 4.e Sequencing 58
- 4.f RNA extraction and reverse transcription 59
- 4.g Quantification of DNA and RNA 61
- 5. AMPLIFICATION OF PLASMID DNA 61
  - 5.a Transformation of chemically competent E.coli 61
  - 5.b Mini preparation 61
    - 5.b.I. Buffers and solutions 61
    - 5.b.II. Procedure 62
  - 5.c Maxi preparation 63
    - 5.c.I. Buffers and solutions 63
    - 5.c.II. Procedure 63
- 6. ANALYSIS OF PROTEINS 65
  - 6.a SDS-PAGE 65
    - 6.a.I. Buffers and solutions 65
    - 6.a.II. Gel preparation and electrophoresis of proteins 66
  - 6.b Staining of proteins with Coomassie 67
    - 6.b.I. Buffers and solutions 67
    - 6.b.II. Procedure 67
  - 6.c Staining of proteins with silver 68
    - 6.c.I. Buffers and solutions 68
    - 6.c.II. Procedure 68
  - 6.d Western blot 69
    - 6.d.I. Buffers and solutions 69
    - 6.d.II. Blotting 70
    - 6.d.III. Immunodetection 71
  - 6.f Analysis of radioactively labeled proteins and probes 71
  - 6.g Quantification of proteins 72
    - 6.g.I. Bradford assay 72
    - 6.g.II. Semi-quantitative SDS-PAGE 72
    - 6.g.III. Quantification of proteins in small volumes 73
- 7. PURIFICATION OF GST-FUSION PROTEINS 73
  - 7.a Screening 73
  - 7.b Large scale purification 74

7.b.I. Buffers and solutions 74
7.b.II. Preparation of glutathione agarose 75
7.b.III. Purification by affinity chromatography 75
7.c Elution of GST-fusion proteins from the agarose 76
7.c.I. Buffers and solutions 76
7.c.II. Procedure 77
8. [^{35}S]-LABELING OF PROTEINS 77
9. GST PULL-DOWN ASSAY 78
9.a Buffers and solutions 78
9.b Assay conditions 79
10. CROSSLINKING OF PROTEINS USING GLUTARALDEHYDE 80
10.a Buffers and solutions 80
10.b Reaction conditions 81
10.b.I. Crosslinking of GST-TORC1$_{1-44}$ 81
10.b.II. Crosslinking of [^{35}S]TORC1$_{327}$ 82
11. ELECTROPHORETIC MOBILITY SHIFT ASSAY 82
11.a Buffers and solutions 82
11.b Gel preparation 83
11.c Labeling of oligonucleotides 84
11.c.I. Annealing 84
11.c.II. Labeling with [^{32}P] 84
11.c.III. Purification of the labeled probe 84
11.c.IV. Quantification of the incorporation 85
11.d Binding reaction 85
11.d.I. General procedure 85
11.d.II. Supershift 85
11.e Electrophoresis 86
12. TRANSIENT TRANSFECTION OF HIT-T15 CELLS 87
12.a Buffers and solution 87
12.b Transfection using DEAE Dextran 87
12.c Transfection using Metafectene 88
13. TREATMENT OF HIT-T15 CELLS 89
14. PREPARATION OF WHOLE-CELL EXTRACTS (HOT LYSIS) 90
15. PREPARATION OF CYTOSOLIC AND NUCLEAR EXTRACTS 90

Contents

    15.a Buffers and solutions 90
    15.b Extraction 91
  16. IMMUNOCYTOCHEMISTRY 92
    16.a Buffers and solutions 92
    16.b Staining procedure 92
    16.c Fluorescence microscopy 93
  17. CHROMATIN-IMMUNOPRECIPITATION (CHIP) 94
    17.a Buffers and solutions 94
    17.b ChIP 96
    17.c Quantitative real-time PCR 98
  18. LUCIFERASE REPORTER-GENE ASSAY 99
    18.a Buffers and solutions 99
    18.b Preparation of cell extracts 100
    18.c Determination of luciferase activity 101
    18.d Measurement of GFPtpz fluorescence 101
  19. STATISTICS 102

RESULTS **103**

  1. EXPRESSION AND PURIFICATION OF GST-FUSION PROTEINS 103
  2. IN VITRO RADIOACTIVE LABELING OF PROTEINS 105
  3. NUCLEAR TRANSLOCATION OF TORC PROTEINS IN HIT-T15 CELLS: EFFECT OF LITHIUM ... 106
    3.a Effects of KCl and Cyclosporin A on the nuclear translocation of TORC proteins 106
    3.b Effects of cAMP and lithium on the nuclear translocation of TORC proteins 108
  4. EFFECTS OF LITHIUM ON THE OLIGOMERIZATION OF TORC1 110
  5. EFFECTS OF LITHIUM ON THE TRANSCRIPTIONAL ACTIVITY OF TORC PROTEINS 112
  6. TRANSCRIPTIONAL ACTIVITY CONFERRED BY TORC1, TORC2, AND TORC3 TO CREB BZIP – EFFECT OF LITHIUM 114
  7. EFFECTS OF LITHIUM ON THE INTERACTION BETWEEN CREB AND TORC1, TORC2, AND TORC3 116
    7.a Mammalian two-hybrid assay 116
    7.b In vitro GST pull-down assay 118
        7.b.I Effects of lithium on the interaction between GST-CREB and full-length [^{35}S]TORC1, [^{35}S]TORC2, or [^{35}S]TORC3 118

7.b.II Effects of lithium on the interaction between GST-CREB-wt and truncated [^{35}S]TORC1$_{327}$, [^{35}S]TORC2$_{347}$, or [^{35}S]TORC3$_{310}$ 120
8. IDENTIFICATION OF TORC ISOFORMS EXPRESSED IN HIT-T15 CELLS 122
9. EXAMINATION OF THE INTERACTION BETWEEN CREB AND TORC1 AT THE SOMATOSTATIN CRE IN THE EMSA 124
10. EFFECTS OF LITHIUM ON THE RECRUITMENT OF TORC1 TO THE PROMOTER 126
11. FURTHER CHARACTERIZATION OF THE INTERACTION BETWEEN CREB AND TORC1 128

11.a Requirement of the first 44 amino acids of TORC1 to confer transcriptional activity to CREB bZip 128

11.b Effect of the R300A mutation of CREB bZip on the interaction between CREB and TORC1 as revealed by the mammalian two-hybrid assay 130

11.c In vitro GST pull-down assay 132

11.c.I Concentration-response curve of the effect of lithium on the specific interaction between GST-CREB and [^{35}S]TORC1 132

11.c.II Effect of magnesium on the interaction between GST-CREB and [^{35}S]TORC1 134

12. MUTATION OF CREB AT LYSINE 290 (K290) 136
13. DNA-BINDING OF CREB-K290E 137
14. EFFECT OF K290E AND K290A MUTATIONS ON CREB TRANSCRIPTIONAL ACTIVITIES 139

14.a Effect of the K290E mutation on CREB transcriptional activity under basal conditions and after stimulation by KCl and forskolin 139

14.b Effect of K290E and K290A mutations on basal CREB transcriptional activity and on its stimulation by lithium in the presence of cAMP 141

14.c Comparison of the expression level of GAL4-CREB wild-type and K290E/K290A mutants 143

14.d Effect of K290E and K290A mutations on the transcriptional activity conferred by TORC1 to the CREB bZip 144

14.d.I Effect of K290E and K290A mutations on the transcriptional activity conferred by TORC1 to the CREB bZip and its stimulation by lithium in the presence of cAMP 144

14.d.II Specificity of the effect of TORC1 overexpression on GAL4-bZip transcriptional activity 146

15. EFFECT OF K290E AND K290A MUTATIONS ON THE INTERACTION OF CREB WITH TORC1 AND ITS STIMULATION BY LITHIUM 148

- 15.a Mammalian two-hybrid assay 148
- 15.b In vitro GST pull-down assay 152
  - 15.b.I Effect of lithium 152
  - 15.b.II Effect of magnesium 154
- 16. EFFECT OF K290E AND K290A MUTATIONS ON THE RECRUITMENT OF TORC1 TO CREB BZIP AT THE PROMOTER 156
- 17. STIMULATION BY LITHIUM OF CREB/TORC-DIRECTED GENE TRANSCRIPTION INDUCED BY cAMP – EVIDENCE AT NATIVE HUMAN PROMOTERS 158
  - 17.a Effects of lithium at the human fos-gene promoter 158
  - 17.b Effects of lithium at the human BDNF(exon IV)-gene promoter 160
  - 17.c Effects of lithium at the human NR4A2-gene promoter 162

DISCUSSION 164

- 1. HIT-T15 CELLS AS A MODEL SYSTEM 164
- 2. CONCENTRATIONS OF LITHIUM USED IN THE PRESENT STUDY 165
- 3. REGULATION OF TORC BY NUCLEAR AND CYTOSOLIC SHUTTLING IN HIT-T15 CELLS 166
- 4. LITHIUM FACILITATES THE OLIGOMERIZATION OF TORC1 166
- 5. LITHIUM DOES NOT INFLUENCE THE TRANSCRIPTIONAL ACTIVITY OF TORC PROTEINS 167
- 6. LITHIUM FACILITATES THE INTERACTION BETWEEN CREB AND TORC PROTEINS 168
- 7. MAGNESIUM INHIBITS THE INTERACTION BETWEEN CREB AND TORC1 170
- 8. THE ROLE OF THE CREB K290 MUTATION FOR THE EFFECT OF LITHIUM ON CREB-TORC1 INTERACTION 170
- 9. LITHIUM ENHANCES THE CAMP-INDUCED CREB-DIRECTED GENE TRANSCRIPTION AT NATIVE HUMAN PROMOTERS 174
  - 9.a The human fos-gene promoter 174
  - 9.b The human BDNF(exon IV)-gene promoter 175
  - 9.c The human NR4A2-gene promoter 176
- 10. CREB AND TORC1 – FUNCTIONAL IMPLICATIONS TO NEUROPLASTICITY 178
- 11. LITHIUM, BIPOLAR DISORDER AND NEUROPLASTICITY 179

SUMMARY AND CONCLUSION 182

REFERENCES 184

ACKNOWLEDGEMENTS 197

APPENDIX A 199

List of Figures

FIGURE 1: ADENYLYL CYCLASE, INOSITOL MONOPHOSPHATASE AND GSK3β ARE DIRECT MOLECULAR TARGETS OF LITHIUM. 21

FIGURE 2: PRIMARY STRUCTURE OF CREB. 24

FIGURE 3: PRIMARY STRUCTURE OF TORC. 26

FIGURE 4: ELEVATED CALCIUM AND CAMP LEVELS LEAD TO THE NUCLEAR TRANSLOCATION OF TORC. 28

FIGURE 5: SCHEMATIC ILLUSTRATION OF SITE-DIRECTED MUTAGENESIS BY PRIMERLESS PCR.....57

FIGURE 6: PRINCIPLE OF QUANTITATIVE REAL-TIME PCR USING TAQMAN™ PROBES. 98

FIGURE 7: RESULTS OF THE EXPRESSION AND PURIFICATION OF GST-FUSION PROTEINS FROM E.COLI. 104

FIGURE 8: SDS-PAGE OF [^{35}S]-LABELED TORC1, TORC2, AND TORC3................... 105

FIGURE 9: NUCLEAR TRANSLOCATION OF ENDOGENOUS TORC PROTEINS IN HIT-T15 CELLS UPON TREATMENT WITH KCL AND CYCLOSPORIN A, ANALYZED BY IMMUNOCYTOCHEMISTRY........ 107

FIGURE 10: NUCLEAR TRANSLOCATION OF ENDOGENOUS TORC IN HIT-T15 CELLS UPON TREATMENT WITH LITHIUM AND CAMP, ANALYZED BY IMMUNOCYTOCHEMISTRY................... 109

FIGURE 11: EFFECTS OF LITHIUM ON THE OLIGOMERIZATION OF ISOLATED TORC1. 111

FIGURE 12: EFFECT OF LITHIUM ON THE TRANSCRIPTIONAL ACTIVITY OF TORC PROTEINS IN LUCIFERASE REPORTER-GENE ASSAYS. 113

FIGURE 13: EFFECTS OF LITHIUM ON THE TRANSCRIPTIONAL ACTIVITY OF GAL4-BZIP UPON OVEREXPRESSION OF TORC1, TORC2, OR TORC3. 115

FIGURE 14: EFFECTS OF LITHIUM ON THE INTERACTION BETWEEN CREB AND TORC1, TORC2, OR TORC3 IN THE MAMMALIAN TWO-HYBRID ASSAY 117

FIGURE 15: EFFECTS OF LITHIUM ON THE INTERACTION BETWEEN CREB AND FULL-LENGTH TORC1, TORC2, OR TORC3 IN THE GST PULL-DOWN ASSAY. 119

FIGURE 16: EFFECTS OF LITHIUM ON THE INTERACTION BETWEEN CREB AND TRUNCATED TORC1$_{327}$, TORC2$_{347}$, OR TORC3$_{310}$ IN THE GST PULL-DOWN ASSAY................... 121

FIGURE 17: EXPRESSION LEVELS OF TORC ISOFORMS IN HIT-T15 CELLS................... 123

FIGURE 18: CREB AND TORC1 INTERACTION AT THE SOMATOSTATIN CRE IN ELECTROPHORETIC MOBILITY SHIFT ASSAYS................... 125

FIGURE 19: EFFECTS OF LITHIUM AND CAMP ON THE RECRUITMENT OF TORC1 TO PROMOTER IN CHROMATIN IMMUNOPRECIPITATION ASSAYS. 127

FIGURE 20: REQUIREMENT OF THE FIRST 44 AMINO ACIDS OF TORC1 TO CONFER TRANSCRIPTIONAL ACTIVITY TO GAL4-BZIP-WT IN LUCIFERASE REPORTER-GENE ASSAYS. 129

FIGURE 21: EFFECTS OF LITHIUM AND CAMP ON THE INTERACTION BETWEEN CREB OR CREB-R300A AND TORC1 IN THE MAMMALIAN TWO-HYBRID ASSAY. 131

FIGURE 22: CONCENTRATION-RESPONSE CURVE OF THE EFFECT OF LITHIUM ON THE SPECIFIC INTERACTION BETWEEN GST-CREB AND [^{35}S]TORC1 IN A GST PULL-DOWN ASSAY. 133

FIGURE 23: EFFECT OF MAGNESIUM ON THE INTERACTION BETWEEN CREB AND TORC1 IN THE GST PULL-DOWN ASSAY. 135

FIGURE 24: MUTATIONS OF THE LYSINE RESIDUE AT POSITION 290 OF CREB. 136

FIGURE 25: EFFECT OF THE MUTATION K290E ON THE DNA-BINDING ABILITY OF CREB. 138

FIGURE 26: EFFECT OF THE K290E MUTATION ON CREB TRANSCRIPTIONAL ACTIVITY UNDER BASAL CONDITIONS AND AFTER STIMULATION BY KCL AND FORSKOLIN. 140

FIGURE 27: EFFECT OF K290E AND K290A MUTATIONS ON BASAL CREB TRANSCRIPTIONAL ACTIVITY AND ON ITS STIMULATION BY LITHIUM IN THE PRESENCE OF CAMP IN LUCIFERASE REPORTER-GENE ASSAYS. 142

FIGURE 28: COMPARISON BY WESTERN BLOT OF THE EXPRESSION LEVELS OF GAL4-CREB-WT AND THE MUTANTS R300A, K290E, AND K290A. 143

FIGURE 29: EFFECTS OF LITHIUM AND CAMP ON THE TRANSCRIPTIONAL ACTIVITY OF THE CREB BZIP CONFERRED BY TORC1 IN LUCIFERASE REPORTER-GENE ASSAYS – COMPARISON BETWEEN CREB BZIP WILD-TYPE AND THE MUTANTS K290E AND K290A. 145

FIGURE 30: TORC1 OVEREXPRESSION SPECIFICALLY ACTIVATES TRANSCRIPTIONAL ACTIVITY OF GAL4-BZIP. 147

FIGURE 31: COMPARISON BY WESTERN BLOT OF THE EXPRESSION LEVELS OF VP16-BZIP WILD-TYPE AND MUTANTS IN HIT-T15 CELLS. 149

FIGURE 32: EFFECT OF K290E AND K290A MUTATIONS ON THE INTERACTION OF TORC1 WITH CREB AND ITS STIMULATION BY LITHIUM AS REVEALED IN A MAMMALIAN TWO-HYBRID ASSAY. 151

FIGURE 33: EFFECT OF K290E AND K290A MUTATIONS ON THE INTERACTION OF TORC1 WITH CREB AND ITS STIMULATION BY LITHIUM AS REVEALED IN THE GST PULL-DOWN ASSAY. 153

FIGURE 34: EFFECT OF K290E MUTATION ON THE INHIBITION BY MAGNESIUM OF THE INTERACTION BETWEEN TORC1 AND CREB IN THE GST PULL-DOWN ASSAY. 155

FIGURE 35: EFFECTS OF LITHIUM ON THE RECRUITMENT OF TORC1 TO GAL4-BZIP AT THE PROMOTER – COMPARISON BETWEEN CREB BZIP WILD-TYPE AND THE MUTANTS K290E AND K290A IN CHROMATIN IMMUNOPRECIPITATION ASSAYS. 157

FIGURE 36: EFFECTS OF LITHIUM AND CAMP ON HUMAN *FOS*-GENE TRANSCRIPTION IN LUCIFERASE REPORTER-GENE ASSAYS. 159
FIGURE 37: EFFECTS OF LITHIUM AND CAMP ON AT THE TRANSCRIPTIONAL ACTIVITY THE HUMAN *BDNF(EXONIV)*-GENE PROMOTER IN LUCIFERASE REPORTER-GENE ASSAYS 161
FIGURE 38: EFFECTS OF LITHIUM AND CAMP ON HUMAN *NR4A2*-GENE TRANSCRIPTION IN LUCIFERASE REPORTER-GENE ASSAYS. 163
FIGURE 39: NOVEL MECHANISM OF LITHIUM ACTION. 181

List of Tables

TABLE 1: EXPRESSION CONSTRUCTS. 43
TABLE 2: LUCIFEREASE REPORTER-GENE CONSTRUCTS. 45
TABLE 3: OVERVIEW OF THE PRIMER PAIRS USED TO GENERATE THE CONSTRUCTS IN THE PRESENT WORK 46
TABLE 4: SYNTHETIC OLIGONUCLEOTIDES USED FOR PCR CLONING OF EXPRESSION CONSTRUCTS AND LUCIFERASE REPORTER-GENE CONSTRUCTS. 47
TABLE 5: SYNTHETIC OLIGONUCLEOTIDES USED FOR SITE-DIRECTED MUTAGENESIS BY PRIMERLESS PCR. 48
TABLE 6: SYNTHETIC OLIGONUCLEOTIDES USED TO SEQUENCE NEWLY GENERATED CONSTRUCTS. 49
TABLE 7: SYNTHETIC OLIGONUCLEOTIDES AND TAQMAN™ PROBES FOR QUANTITATIVE REAL-TIME PCR. 50
TABLE 8: SYNTHETIC OLIGONUCLEOTIDES USED AS DOUBLE-STRANDED PROBES FOR EMSAS. ...50
TABLE 9: APPLICATIONS AND DILUTIONS OF PRIMARY AND SECONDARY ANTIBODIES 53

Abbreviations

aa – amino acids
AC – adenylyl cyclase
$AgNO_3$ – silver nitrate
Amp – ampicillin
AMP – adenosine monophosphate
AMPA – α-amino-3-hydroxy-5-methyl-4-isoxazolepropionic acid
AMPK – AMP-activated protein kinase
ANOVA – analysis of variance
AP1 – activator protein 1
APS – ammonium persulphate
ATP – adenosine triphosphate
BD – bipolar disorder
BDNF – brain-derived neurotrophic factor
BSA – bovine serum albumin
bZip – basic leucine zipper
°C – degree celcius
CaMK – calcium/calmodulin-dependent kinase
cAMP – cyclic adenosine monophosphate
CBP – CREB binding protein
cDNA – complementary DNA
ChIP – chromatin immunoprecipitation
CMV – cytomegalovirus
cpm – counts per minute
CRE – cAMP response element
CREB – cAMP response element binding protein
CREM – cAMP response element modulator
CsA – cyclosporin A
CsCl – cesium chloride
DAG - diacylglycerol
DAPI – 4',6-diamidino-2-phenylindol
dATP – deoxyadenosine triphosphate
dCTP – deoxycytidine triphosphate

dGTP – deoxyguanosine triphosphate
DMSO – dimethyl sulfoxide
DNA – deoxyribonucleic acid
dNTPs – deoxynucleoside triphosphates
DTT – dithiothreitol
dTTP – deoxythymidine triphosphate
EMSA – electrophoretic mobility shift assay
ER – endoplasmatic reticulum
ERK – extracellular signal-regulated kinase
FSK – forskolin
GDP – guanidine diphosphate
GFP – green fluorescent protein
GFPtpz – green fluorescent protein variant topaz
GSK3β – glycogen synthase kinase 3β
GST – glutathione S-transferase
GTP – guanidine triphosphate
h – hour
HCl – hydrochloric acid
IB1 – islet-brain 1
IMPase – inositolmonophosphate phosphatase
IP_3 – inositol triphosphate
IPTG – isopropyl-β-D-thiogalactoside
JIP-1b – JNK-interacting protein 1b
K_2HPO_4 – di-potassium hydrogen phosphate
KCl – potassium chloride
kDa – kilo Dalton
KH_2PO_4 – potassium di-hydrogen phosphate
KID – kinase inducible domain
LiCl – lithium chloride
LTD – long-term depression
LTP – long-term potentiation
MAPK – mitogen-activated protein kinase
MARCKS - myristoylated alanine-rich C kinase substrate
$MgCl_2$ – magnesium chloride

MgSO$_4$ – magnesium sulphate
min - minute
N$_2$ - nitrogen
Na$_2$CO$_3$ – sodium carbonate
Na$_2$HPO$_4$ – di-sodium hydrogen phosphate
Na$_2$S$_2$O$_3$ – sodium thiosulphate
NaAc – sodium acetate trihydrate
NaCl – sodium chloride
NaH$_2$PO$_4$ – sodium di-hydrogen phosphate
NaOH – sodium hydroxide
NES – nuclear export sequence
NLS – nuclear localisation sequence
NMDA – N-methyl-D-aspartate
NONO – non-POU-domain-containing octamer-binding protein
OD – optical density
PBS – phosphate-buffered saline
PCR – polymerase chain reaction
PEG 6000 – polyethylene glycol
PIP$_2$ – phosphatidylinositol 4,5-bisphospate
PKA – protein kinase A
PKC – protein kinase C
PLC – phospholipase C
PMSF – phenylmethylsulfonylfluoride
poly(dI-dC) – poly(deoxyinosinic-deoxycytidylic) acid
RNA – ribonucleic acid
RPMI – Roswell Park Memorial Institute
rpm – rounds per minute
RSK – pp90 ribosomal S6 kinase
SDS – sodium dodecylsulphate
SDS-PAGE – sodium dodecylsulphate polyacrylamide gel electrophoresis
sec – seconds
SEM – standard error of mean
SIK – salt-inducible kinase
somCRE – somatostatin CRE

SSRI – selective serotonin reuptake inhibitor
TORC – transducer of regulated CREB
vol - volume
Wnt – wingless signal
wt – wild-type

Introduction

1. Bipolar disorder

Bipolar disorder (BD) is a severe, chronic and often life-threatening illness. The concept of bipolarity involves the two mood states of mania and depression. Patients typically experience recurrent episodes of mania and depression across their life span, but duration and severity of the episodes greatly vary. The manic state is characterized by euphoria, overactivity, flight of ideas, and positive psychomotor signs. Essentially, the manic behaviour is distinctively different from the patient's usual personality. When untreated a manic phase can last months or years (Belmaker, 2004). In contrast, the depressive state is marked by a depressed mood, anhedonia, suicidal thoughts, and psychomotor slowing. Bipolar I disorder, the classic form of the illness, is defined by the presence of a depressive disorder associated with episodes of mania. Milder forms of mania, without psychotic symptoms and without potentially harmful behaviour to oneself or others, are defined as hypomania. The syndrome of major depressive disorder combined with episodes of hypomania is called bipolar II disorder (Belmaker, 2004). BD occurs worldwide with a life time prevalence of 1% among all populations that have been studied (Escamilla and Zavala, 2008). Regarding the risk to develop BD genetic factors seem to play a role especially in bipolar I disorder, and numerous potential candidate genes have been examined. However, the findings have not yet been verified in human samples large enough to definitely implicate them in the genesis of BD (Escamilla and Zavala, 2008; Kato, 2007; Kato et al., 2007). To pinpoint biological changes underlying the disease, different brain structures have been investigated. Several neuro-imaging studies were performed to examine putative structural changes in brain regions involved in the regulation of mood and behaviour (Adler et al., 2006; Scherk et al., 2004; Zarate et al., 2003). Dysfunctional changes have been found in an anterior limbic network including portions of the frontal cortex and the cerebellum. The prefrontal cortex is a brain region integrating stimulus-reward associations, reward-guided behaviour and the modulation of emotion, as well as it contributes to attention and short-term memory. Indeed, morphometric abnormalities have been identified in patients with BD in the prefrontal cortex marked by a decrease in volume and changes in functional activity. Especially, the orbitofrontal cortex showed a bilateral volume reduction. With regard to the limbic system, a brain structure involved in emotion, behaviour and long-term memory, increased volume of the amygdala and reduced volume of the hippocampus was reported (Adler et al., 2006; Frey et al., 2007; Scherk et al., 2004). The cerebellum is a structure only recently identified to be involved in cognition and affect (Baldacara et al., 2008). Neuroanatomical studies revealed a fronto-cerebellar connectivity, consisting of closed cortico-cerebellar loops in which the prefrontal cortex connects to the cerebellum via pontine nuclei while the cerebellum sends projections back to the prefrontal cortex via the dentate nucleus and thalamus (Baillieux et al., 2008). Indeed, patients with BD showed a reduced volume in the cerebellum compared to healthy volunteers (Adler et al., 2006; Baldacara et al., 2008; Scherk et al., 2004). Albeit structural changes of the brain have been associated to bipolar disorder the pathogenesis of the illness remains unclear.

To treat acute mania and for the prophylaxis of recurrent episodes, lithium is the medication of choice (Vieta and Sanchez-Moreno, 2008). At optimal dosing it reduces 50% of the recurrences (Vieta and Sanchez-Moreno, 2008). Moreover, substantial antisuicidal effects of lithium have been reported (Maj, 2003).

2. Pharmacology of lithium

2.a Pharmacokinetic of lithium

Lithium's efficacy to treat mania was initially described 60 years ago (Cade, 1949). As a drug it has a very narrow therapeutic range with recommended plasma levels between 0.5 and 1.2 mmol/L, whereas levels above 2 mmol/L are considered as toxic (Mota de Freitas et al., 2006). Lithium is administered orally and is absorbed readily from the gastro intestinal tract by passive diffusion processes (Birch, 1999). Peak blood levels are reached within 30 min to 3 h. It has a plasma half life of circa 24 h; it is not metabolized but excreted by the kidney. 80% of the filtered lithium is reabsorbed by the proximal renal tubulus due to a competition of lithium with sodium in the sodium-proton antiporter. Therefore the loss of sodium, for instance by heavy sweating, enhances the reabsorption of lithium which is clinically important with respect to its narrow therapeutic range (Baldessarini and Tarazi, 2006).

Different pathways of lithium transport across cell membranes have been examined *in vitro* (Birch, 1999). Lithium can replace potassium extracellularly at the sodium-potassium ATPase and is thereby transported into the cell; but lithium can also replace sodium intracellularly and is transported out of the cell. At the chloride-dependent sodium-potassium cotransport system lithium is thought to replace sodium. Lithium also undergoes anion exchange in a cotransport with bicarbonate. The divalent carbonate ion is capable of forming negatively charged ion pairs with sodium or lithium gaining access to the anion exchange system. The single charged ion pair $(Li^+ + CO2^{2-})^-$ exchanges for a monovalent anion such as chloride. Lithium efflux occurs via a sodium-lithium countertransport. Lithium substitutes for sodium in the ATP-independent Na^+-Na^+ countertransport. Lithium also undergoes a downhill transport called leak which is probably shared by sodium and potassium (Birch, 1999). Montezinho and coworkers described the uptake of lithium by the Na^+/Ca^{2+} antiporter most likely in exchange with intracellular Ca^{2+} in human neuroblastoma SH-SY5Y cells (Montezinho et al., 2004). The lithium-sodium countertransport, anion exchange and the leak mechanism are considered to be predominant *in vivo*. All are potentially bidirectional, but the overall direction of flux under physiological conditions is efflux, supporting the view that the cellular uptake of lithium is very low. However, by now the *in vivo* intracellular concentration of lithium in excitable cells like neurons could not be determined (Birch, 1999).

2.b Pharmacodynamic of lithium

Lithium is the lightest of the alkali metals. It shares chemical properties with magnesium due to similar electronegativities and Pauling ionic radii of their cations with 0.60 Å for Li^+ and 0.65 Å for Mg^{2+} being in the same range of magnitude. Both share comparable selectivity for ligand-binding, resulting in the potential interference of lithium with magnesium-dependent cellular processes (Birch, 1999; Mota de Freitas et al., 2006; Quiroz et al., 2004). Several molecular targets of lithium have been described. Effects of lithium on the signal transduction mediated by cyclic adenosine monophosphate (cAMP), phosphatidylinositol, and glycogen synthase kinase 3β (GSK3β) are the longest-known effects and presented here as examples.

2.b.I Cyclic AMP

Lithium interferes with the signalling cascade mediated by the second messenger cyclic adenosine monophosphate (cAMP) by a mechanism not completely understood so far (Quiroz et al., 2004). This cascade involves receptors coupled to a heterotrimeric guanine nucleotide-binding (G) protein. Various receptors of neurotransmitters, like serotonin, dopamine, or noradrenaline for instance, are coupled to G-proteins. The G-protein consists of 3 subunits α, β, and γ. Upon ligand binding to the receptor the α subunit of the G-protein exchanges GDP for GTP and becomes activated. The α subunit dissociates from the β/γ complex and activates or inhibits the enzyme adenylyl cyclase (AC), depending on the subtype of the α subunit α_s or α_i, respectively. Guanosine triphosphate (GTP) is hydrolyzed to guanosine diphosphate (GDP) and the α subunit reassociates to the β/γ subunit (Lodish et al., 2004). The enzyme AC catalyzes the formation of cAMP from AMP, which activates protein kinase A (PKA). cAMP binds to the regulatory subunits of PKA which leads to the release of the catalytic subunits thereby activating the kinase activity of PKA (Lodish et al., 2004). Lithium exerts complex influence on this second messenger cascade (Figure 1). In vitro data support an inhibitory effect of lithium directly on the catalytic subunit of the AC. The effect of lithium is possibly due to a competition with magnesium as the effect was overcome by increasing magnesium concentrations (Newman and Belmaker, 1987). Consistent with this inhibition, lithium was reported to reduce stimulated cAMP levels induced for instance by forskolin, a potent activator of the AC (Chen et al., 1999; Quiroz et al., 2004). Noteworthy, it is thought that under basal conditions cAMP production is tonically inhibited by predominant influence of α_i (Jope, 1999). Here, as another point of action, a direct inhibition of the α_i subunit of the G-protein by lithium was reported. Probably also due to a competition with magnesium lithium interferes with the binding of GTP to α_i (Mota de Freitas et al., 2006). Thus, it was proposed that lithium shifts the equilibrium of the free active conformation of α_i and its inactive conformation associated to β/γ towards the inactive form. Consequently the prevailing tonic inhibition by α_i is reduced thereby increasing basal cAMP levels (Jope, 1999; Mota de Freitas et al., 2006; Quiroz et al., 2004). These rather opposing effects of lithium were proposed to contribute jointly to stabilization of this signalling cascade by minimizing maximal fluctuation of the signalling by cAMP, with lithium elevating minimal levels and attenuating peak levels (Jope, 1999).

2.b.II Phosphatidylinositol

G-protein coupled receptors coupled to the α_q subunit activate the phosphoinositide pathway. The membrane associated phospholipase C (PLC) hydrolyzes the phosphatidylinositol 4,5-bisphospate (PIP_2) to generate two important second messengers, inositol 1,4,5-triphosphate (IP_3) and diacylglcerol (DAG). IP_3 diffuses in the cytosol and triggers the release of Ca^{2+} from ER lumen. Increased intracellular Ca2+ levels activate for instance calcium calmodulin dependent kinases (CaMK) or trigger the synaptic release of vesicles in neuronal cells. DAG activates protein kinase C (PKC) which in turn regulates numerous other proteins (Lodish et al., 2004). The recycling of IP_3 involves the inositolmonophosphate phosphatase (IMPase) as rate limiting enzyme in the conversion of IP_3 to inositol as precursor of PIP_2. In this context, lithium was shown to directly inhibit IMPase (Figure 1) (Chen et al., 1999; Mota de Freitas et al., 2006; Quiroz et al., 2004). This fact led to the hypothesis that the action of lithium is due to the reduction of free inositol in the brain. Several studies have investigated the so-called "inositol depletion hypothesis" but up to now the results are conflicting (Chen et al., 1999; Mota de Freitas et al., 2006; Quiroz et al., 2004).

2.b.III Glycogen synthase kinase 3β

Glycogen synthase kinase 3β (GSK3β) is a serine-threonine kinase which is normally highly active in cells. Signals arising from several pathways including PKA, PKC, or the wingless (Wnt) pathway for instance, lead to the deactivation of GSK3β, thereby influencing many other transduction pathways. Numerous targets of GSK3β are known and include transcription factors (like cAMP-response element binding protein [CREB], β-catenin, or c-Jun), proteins bound to microtubules (like Tau or kinesin light chain), or regulators of cell metabolism (like glycogen synthase or pyruvate dehydrogenase) (Quiroz et al., 2004). Lithium was shown to directly inhibit GSK3β by competition with Mg^{2+} at the non-ATP:Mg^{2+} magnesium-binding site of GSK3β thereby interfering with signalling cascades involving GSK3β (Figure 1). This inhibition was present at therapeutically relevant concentrations of 1-2 mM (Ryves et al., 2002; Ryves and Harwood, 2001). Indeed, GSK3β is known to be crucial in the central nervous system. For instance, GSK3β was found to be the key kinase phosphorylating Tau in neurons. Tau is a microtubule-associated protein that binds to tubulin to promote microtubule assembly.

Hyperphosphorylated Tau is thought to result in destabilization of microtubules and subsequently in loss of dendritic microtubules and cytoskeletal degeneration. Of note, GSK3β is phosphorylating the majority of sites of Tau that are abnormally phosphorylated in brains of patient with Alzheimer's disease (Bhat et al., 2004). Moreover, GSK3β influences transcriptional regulation by phosphorylation of c-Jun and β-catenin (Chen et al., 1999). In addition, the transcription factor CREB was found to be phosphorylated by GSK3β at serine 115 (Fiol et al., 1994). The function of this event is still not clear, but it may enhance the activity of CREB at certain gene promoters and reduce the activity at others (McClung and Nestler, 2008).

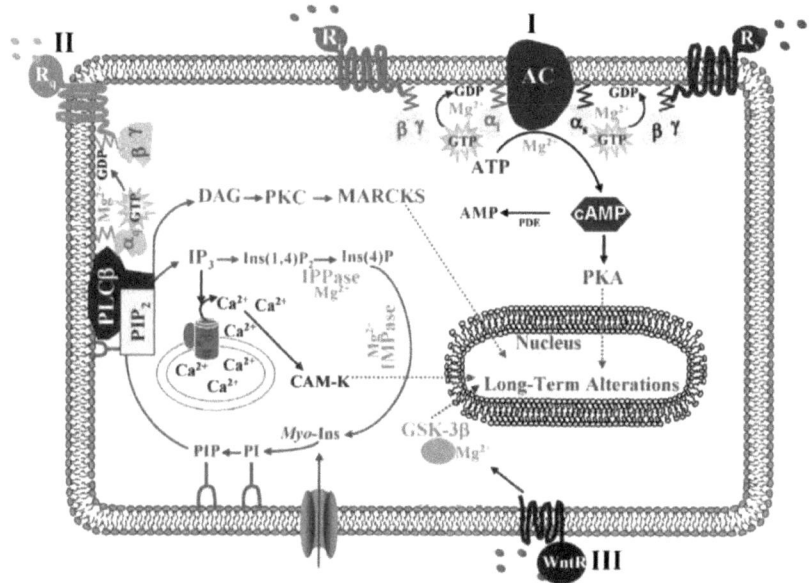

Figure 1: Adenylyl cyclase, inositol monophosphatase and GSK3β are direct molecular targets of lithium.

Pathway I: Membrane bound G-protein coupled receptors R_i and R_s, inhibit or activate the adenylyl cyclase (AC), respectively, depending on the type of the α subunit of the heterotrimeric G-protein. AC catalyzes the formation of cyclic adenosine monophosphate (cAMP) which activates protein kinase A (PKA). PKA can phosphorylate many other proteins. Lithium reduces cAMP levels due to an inhibition of the AC by a competition with magnesium at magnesium binding sites of the catalytic subunit of the AC. By competing with magnesium lithium interferes with the binding of guanosine triphosphate (GTP) to the α subunit of the G-protein thereby reducing the respective inhibitory or activating effects elicited by the GTP-activated α subunit.

Pathway II: Membrane bound G-protein coupled receptors R_q activate phospholipase C (PLC) which hydrolyzes phosphatidylinositol 4,5-bisphospate (PIP_2) to inositol 1,4,5-triphosphate (IP_3) and diacylglcerol (DAG). DAG activates PKC which in turn phosphorylates numerous target proteins, for instance myristoylated alanine-rich C kinase substrate (MARCKS). IP_3 induces Ca^{2+} release from intracellular stores which activate calcium/calmodulin-dependent kinases (CaMK). Lithium inhibits the inositol monophosphatase (IMPase), the enzyme mediating in the rate limiting step of the conversion of IP_3 to insositol, the precursor of PIP_2.

Pathway III: Binding of the wingless signal to its receptor (WntR) reduces GSK3β activity which in turn leads for instance to the accumulation of β-catenin in the nucleus accompanied by changes in transcription. Lithium inhibits GSK3b by competition with magnesium at the non-ATP:Mg^{2+} magnesium binding site. The figure was modified from Mota de Freitas et al., 2006.

Although the effectiveness of lithium in the treatment of BD is indisputable and various direct and downstream molecular targets of lithium have been identified, the mechanism of action underlying the clinical effect of lithium is still not clearly understood. The inhibition of enzymes and the subsequent effect on signalling processes are believed to contribute some beneficial effects in BD (Mota de Freitas et al., 2006; Phiel and Klein, 2001; Quiroz et al., 2004). However, the therapeutic concept of lithium treatment requires the chronic administration. To reach the maximal clinical efficacy treatment for several weeks up to six months has been shown to be necessary. Importantly, the changes induced by enzyme inhibition are instantaneous and are therefore not sufficient to explain the delay until onset of clinical efficacy (Mota de Freitas et al., 2006; Phiel and Klein, 2001; Quiroz et al., 2004). This indicates, besides immediate enzymatic effects, the contribution of other rather long lasting processes. In particular, changes in gene expression accompanied by neuronal adaptation are suggested to be involved (Quiroz et al., 2004; Zarate et al., 2003). The best-studied transcription factor in the context of neuronal adaptation is the cAMP response element binding protein CREB (Lonze and Ginty, 2002; McClung and Nestler, 2008), which has been implicated before to mediate long-term changes induced by lithium treatment (Chuang et al., 2002; Manji et al., 2000).

3. The transcription factor CREB

The cAMP response element binding protein CREB is a ubiquitously expressed transcription factor. It has been found to play a pivotal role in various physiological and developmental processes like learning and memory (Carlezon et al., 2005), glucose homeostasis (Herzig et al., 2003), cell survival (Mayr and Montminy, 2001) as well as neuronal development (Mantamadiotis et al., 2002). In accordance with that, CREB regulates the transcription of genes encoding for instance dopamine β-hydroxylase, somatostatin, glucagon, insulin, other transcription factors (e.g. c-Fos, Nurr1), or growth factors (e.g. brain-derived neurotrophic factor), among many others (Mayr and Montminy, 2001). Mantamadiotis and coworkers investigated effects of the loss of CREB on the development of the brain by generation of homozygous CREB-knockout ($CREB^{-/-}$) mice. $CREB^{-/-}$ mice died perinatally, but the loss of CREB was accompanied by an upregulation of the cAMP response element modulator (CREM), a member of the CREB family which potentially compensated in part for the loss of CREB. The conditional disruption of CREB-

and CREM-function in the brain during development caused the perinatal death of mice due to generalized cell death in the nervous system. Strikingly, the postnatal knockout of CREB and CREM induced progressive neurodegeneration in the dorsolateral striatum as well as in the CA1 and dentate gyrus region of the hippocampus (Mantamadiotis et al., 2002). These facts underline the importance of CREB for the development and maintenance of neuronal function.

3.a The structure of CREB

The initial cloning of the CREB cDNA revealed α-CREB and Δ-CREB, or $CREB_{341}$ and $CREB_{327}$, respectively. Resulting from alternative splicing of one of the eleven exons of the *Creb* gene the two forms differ in the presence of a 14 amino acid stretch termed the α-peptide (Figure 2), but are functionally equal. By alternative splicing of several 5' exons the isoform β-CREB is generated: a protein lacking the first 40 amino acids and the α-peptide compared to α-CREB (Figure 2). All forms are expressed uniformly in human somatic cells (Lonze and Ginty, 2002; Shaywitz and Greenberg, 1999). The amino acid numbering in the present work refers to the Δ-CREB isoform.

CREB, as well as its family members activating transcription factor 1 (ATF1) and cAMP-response element modulator (CREM), belongs to the family of basic leucine zipper (bZip) transcription factors (Mayr and Montminy, 2001). The primary structure of CREB, as shown in Figure 2, demonstrates a centrally located 60 amino acid stretch referred to as the kinase inducible domain (KID). The hydrophobic glutamine-rich domains Q1 and Q2 flank the KID and are considered to be constitutively active. Dimerization of two CREB monomers is mediated by a conserved heptad repeat of leucine residues at the C-terminus called the leucine zipper. N-terminal to the leucine zipper, a lysine- and arginine-rich basic domain conveys the binding of CREB to DNA containing the octamer core sequence 5'-TGACGTCA-3', the cAMP response element (CRE) (Lonze and Ginty, 2002; Mayr and Montminy, 2001). By means of x-ray christallography the structure of the bZip bound to the CRE of the *somatostatin*-gene promoter was elucidated by Schumacher and colleagues in 2000. The analysis revealed the formation of a continuous α-helix by the bZip, in which the leucine zipper region forms a parallel coiled-coil interaction interface, and the basic region contacted the major groove of the DNA (Schumacher et al., 2000). Interestingly, the crystal structure revealed a hexahydrated magnesium ion in the cavity between the bifurcating basic regions. The magnesium ion is located between the extended side-chains of the lysine-290 residues (K290) of the homodimer (Schumacher et al., 2000). The DNA binding of CREB strongly depends on the ability to coordinate that magnesium ion which was shown to be disrupted upon mutation of K290 (Craig et al., 2001; Dwarki et al., 1990; Schumacher et al., 2000).

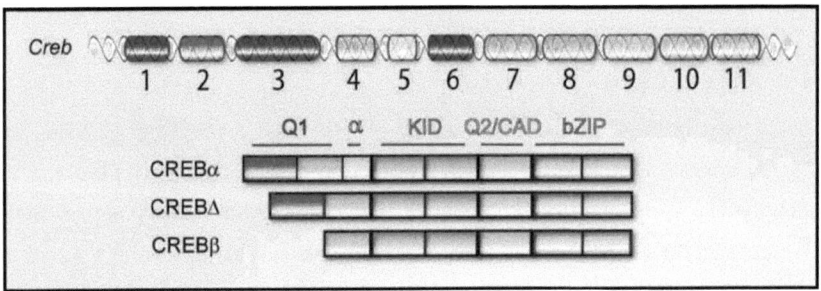

Figure 2: Primary structure of CREB.
The *Creb* gene is composed of eleven exons (top). The α-CREB contains a C-terminal basic leucine zipper (bZip). The transactivation domain is composed of the central kinase inducible domain (KID) flanked by glutamine-rich domains Q1 and Q2, which are considered to be constitutively active (CAD). Upon alternative splicing Δ-CREB is generated lacking a 14 amino acid stretch termed α-peptide in the α-CREB isoform. Alternative splicing of several 5' exons generates β-CREB lacking the first 40 amino acids of α-CREB and the α-peptide. The figure was modified from Lonze and Ginty, 2002.

3.b Transcriptional regulation mediated by CREB

The CREB-directed gene transcription is distinctly induced in response to numerous different signalling pathways. One critical trait is the phosphorylation of serine 119 (S119) situated in the KID (Mayr and Montminy, 2001). For instance, PKA phosphorylates CREB at S119 in response to elevated cAMP levels (Figure 4), and calcium/calmodulin-dependent kinases I, II and IV (CaMK I, II, and IV) phosphorylate S119 upon increased intracellular calcium levels after membrane depolarization (Figure 4). Furthermore, growth factors also lead to the phosphorylation of CREB by activation of pathways involving the family of mitogen-activated protein kinases (MAPK) resulting in the phosphorylation of S119 by the pp90 ribosomal S6 kinase family (RSKs) (Carlezon et al., 2005; Mayr and Montminy, 2001; Shaywitz and Greenberg, 1999). Essentially, the interaction with co-factors is a key feature of transcription regulation (Johannessen et al., 2004). CREB can interact with at least 30 other proteins affecting the CREB-directed gene transcription

(Johannessen et al., 2004; McClung and Nestler, 2008). One of the best-described coactivators of CREB is the CREB binding protein (CBP) which is recruited to CREB in response to phosphorylation of S119. CBP possesses intrinsic histone deacetylase activity and is thought to promote CREB-directed transcription by association with RNA-polymerase II complexes (Mayr and Montminy, 2001; Shaywitz and Greenberg, 1999). Additionally, CREB-directed gene transcription is promoted by interaction of the CREB-Q2 domain with $TAF_{II}130$ of the TFIID complex belonging to the general transcriptional machinery (Nakajima et al., 1997).
Although the phosphorylation of CREB at S119 is believed to be necessary to activate CREB-directed gene-transcription, it is not sufficient. The immunosuppressive drugs cyclosporin A and FK506 potently block the stimulated CREB-directed gene transcription without interfering with the phosphorylation of S119 (Oetjen et al., 2005; Schwaninger et al., 1995; Schwaninger et al., 1993a). In 2003 Iourgenko et al. identified a new coactivator of CREB that associates to the bZip and promotes CREB-directed gene transcription independent of phosphorylation at S119. This new co-activator of CREB is termed transducer of regulated CREB (TORC) (Iourgenko et al., 2003).

4. The CREB coactivator TORC

A library of putative full-length human cDNA clones was searched by high throughput sreening for activators of the *interleukin 8* gene promoter and identified the transducer of regulated CREB 1 (TORC1) as coactivator of the transcription factor CREB (Iourgenko et al., 2003). Database searches identified TORC2 and TORC3 to share 32% identity with TORC1 (Iourgenko et al., 2003). TORC1, TORC2, and TORC3 orthologs were identified in mice, whereas TORC1 orthologs are also present in fugu and drosophila (Iourgenko et al., 2003). TORC proteins are expressed at low levels in most tissues but the expression pattern differs among the isoforms. TORC1 is predominantly expressed in brain tissue, especially in the prefrontal cortex and cerebellum, whereas TORC2 and TORC3 were predominantly found in B- and T-lymphocytes (Conkright et al., 2003a). Functionally, TORC1 has been implicated to be involved in the maintenance of the late phase of long-term potentiation in the hippocampus (Kovacs et al., 2007; Zhou et al., 2006). In contrast, TORC2 seems to be rather involved in the regulation of glucose homeostasis (Dentin et al., 2008; Dentin et al., 2007; Koo et al., 2005; Liu et al., 2008; Screaton et al., 2004).

4.a The structure of TORC

The primary structure of TORC proteins reveals a highly conserved N-terminal predicted coiled-coil domain (Figure 3). This domain is mediating the interaction with CREB independent of the phosphorylation of CREB at S119. Rather the amino acid arginine 300 (R300) of the CREB bZip has been shown to be indispensable for the interaction with TORC proteins (Conkright et al., 2003a; Iourgenko et al., 2003). By means of glutaraldehyde cross-linking, TORC was found to oligomerize and was suggested to bind to the bZip as a tetramer (Conkright et al., 2003a). All human TORC isoforms possess a potent transactivation domain at the C-terminus (Figure 3) able to induce a minimal promoter in luciferase reporter gene assays (Iourgenko et al., 2003). Additionally, TORC1 and TORC2 were shown to potentiate the CREB-directed gene transcription in response to elevated Ca^{2+}- and cAMP-levels (Conkright et al., 2003a; Screaton et al., 2004). Analysis of the sequence of TORC2 revealed the presence of a nuclear localization sequence (NLS) at amino acids 56-144 and two nuclear export sequences (NES) in the amino acid region 145-320 (Figure 3). The mutation of the specific residues in NES1 and NES2 lead to nuclear accumulation of TORC2, arguing for the functional relevance of these sites. Moreover, the NLS and NES1/2 are conserved among the three isoforms (Screaton et al., 2004).

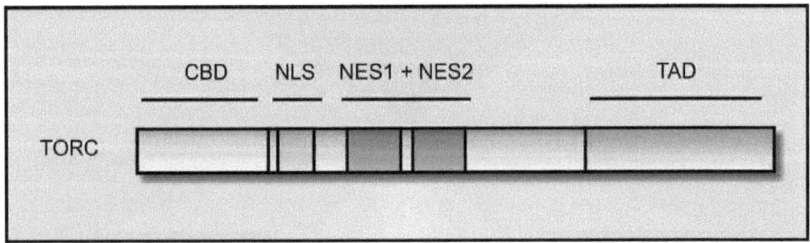

Figure 3: Primary structure of TORC.
At the N-terminus TORC contains a highly conserved predicted coiled-coil domain. This domain mediates the interaction with CREB (CBD; CREB binding domain). C-terminal to the CBD, the nuclear localisation signal (NLS) and two nuclear export sequences (NES) are located. C-terminally, TORC contains a transactivation domain (TAD). The figure was modified from Screaton et al., 2004.

4.b Regulation of TORC

The potentiation of CREB-directed gene transcription by TORC is tightly regulated by nuclear and cytosolic shuttling of TORC proteins. Under resting conditions, TORC is sequestered in the cytoplasm bound by 14-3-3 phospho-protein binding proteins (Screaton et al., 2004). Salt inducible kinase (SIK), a member of the family of AMP-activated protein kinases (AMPK), was identified to phosphorylate TORC proteins leading to their cytosolic retention (Figure 4, I.) (Katoh et al., 2004). Upon elevated levels of intracellular cAMP or Ca^{2+}, signals enhancing also CREB-directed gene transcription (Figure 4), TORC proteins translocate into the nucleus where they can interact with the bZip of CREB (Bittinger et al., 2004; Screaton et al., 2004). Elevated levels of cAMP activate PKA which was shown to phosphorylate SIK at serine 577 (S577) thereby inhibiting the kinase activity of SIK (Figure 4, II.) (Takemori and Okamoto, 2008). In contrast, increases in intracellular Ca^{2+} levels activate the calcium/calmodulin-dependent phosphatase calcineurin directly dephosphorylating TORC proteins (Figure 4, III.) which leads to their nuclear accumulation (Bittinger et al., 2004; Screaton et al., 2004). Though the nuclear translocation of TORC is necessary and sufficient to potentiate CREB-directed transcription independent from the phosphorylation of CREB at S119 (Bittinger et al., 2004), recent studies suggested that TORC regulates CREB-directed transcription in cooperation with CBP (Ravnskjaer et al., 2007; Xu et al., 2007).

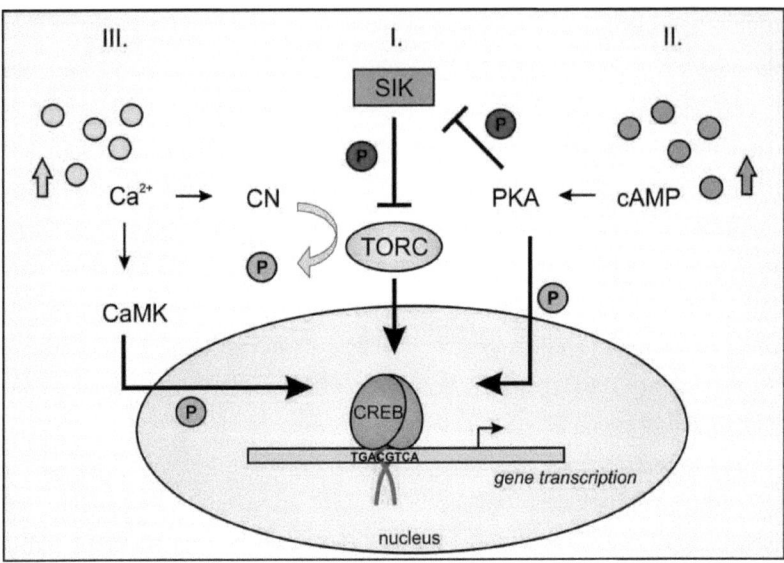

Figure 4: Elevated calcium and cAMP levels lead to the nuclear translocation of TORC.

I. Under resting conditions TORC proteins are phosphorylated by salt inducible kinase (SIK). The phosphorylation leads to the cytosolic retention of TORC being bound by 14-3-3 proteins (not shown).
II. Elevated levels of cAMP activate protein kinase A (PKA). PKA phosphorylates CREB at S119 activating CREB-directed transcription. In addition, PKA phosphorylates SIK at S577 inhibiting the phosphorylation of TORC by SIK.
III. Upon elevation of intracellular Ca^{2+} levels the calcium/calmodulin-dependent kinase (CaMK) is activated. CaMK phosphorylates CREB at S119 activating CREB-directed transcription. The calcium/calmodulin dependent phosphatase calcineurin (CN) is also activated by increased Ca^{2+}. CN dephosphorylates TORC leading to the nuclear accumulation of TORC where it can bind to the bZip of CREB.

5. Lithium enhances cAMP-induced CREB-directed gene transcription

Recently, lithium was shown to enhance cAMP-induced CREB-directed gene transcription in cell cultures using the membrane-depolarizable β-cell line HIT-T15 (Boer et al., 2007). This effect was concentration-dependent and not due to an inhibition of GSK3β or depletion of inositol. Moreover, neither CREB phosphorylation nor CREB DNA-binding was affected by lithium. Instead the presence of the bZip was indispensable for the enhancement by lithium of cAMP-induced CREB-directed gene transcription suggesting mediation by TORC (Boer et al., 2007). Indeed, TORC was identified to confer this effect of lithium since the overexpression of TORC1 restored transcriptional activity of CREB bZip in response to cAMP and lithium (Boer et al., 2007).

Thus, Boer and colleagues provided evidence that lithium directly affects CRE/CREB-directed gene transcription and, moreover, identified the CREB coactivator TORC as novel target of lithium action. This effect might play a role in lithium-induced neuronal adaptation and contribute to the clinical efficacy of lithium in the treatment of BD.

6. Objectives of the study

The present study aimed to elucidate the molecular mechanism by which lithium through TORC stimulates cAMP-induced CREB-directed transcription.

For that purpose effects of lithium on TORC itself have been examined with respect to the nuclear accumulation, the transcriptional potential, the oligomerization, and the recruitment to the promoter. Moreover, the interaction between CREB and TORC was characterized in cell-free conditions and in cell culture. Due to the potential of lithium to interfere with magnesium-binding in different enzymes and the fact that CREB complexes a magnesium ion in the bZip, the putative contribution of a competition between lithium and magnesium was investigated. In conjunction to that, the role of the amino acid lysine 290 of CREB for the enhancement by lithium of the cAMP-induced CREB-directed transcription was characterized. Additionally, the three different isoforms of TORC were compared with respect to their ability to mediate the effects of lithium. Finally, the effects of lithium on cAMP-induced CREB-directed gene transcription mediated by TORC were demonstrated at native human gene promoters.

Materials and methods

1. Equipment

1.a Apparatus

ABI PRISM 3100 Genetic Analyzer – *Applied Biosystems, Darmstadt, Germany*
ABI PRISM 7900 HT Sequence Detection System – *Applied Biosystems, Darmstadt, Germany*
AutoLumat LB 953 luminometer – *E&G Berthold, Bad Wildbach*
Bacteria incubator 37°C – *Heraeus Sepatech, Langenselbold, Germany*
BAS-MS 2325 phosphor-imager screen – *FUJIFILM, purchased by raytest Isotopenmessgeräte GmbH, Straubenhardt, Germany*
BAS-1800II phosphor-imaging device – *FUJIFILM, purchased by raytest Isotopenmessgeräte GmbH, Straubenhardt, Germany*
Beckmann centrifuge GS-6 – *Beckmann GmbH, München, Germany*
Beckmann J2HS centrifuge – *Beckmann GmbH, München, Germany*
Beckmann L8-70M Ultracentrifuge – *Beckmann GmbH, München, Germany*
Beckmann Tube Sealer – *Beckmann GmbH, München, Germany*
Beta counter LS1801 – *Beckmann GmbH, München, Germany*
Biofuge 15R – *Heraeus Sepatech, Langenselbold, Germany*
Biofuge pico – *Heraeus Sepatech, Langenselbold, Germany*
Biometra® Standard Power Pack P25 – *Biometra, Göttingen, Germany*
Blot chamber – *Amersham Biosciences, Freiburg, Germany*
Branson Sonifyer® Cell Disrupter B15 – *Heinemann Ultraschall- und Labortechnik, Schwäbisch Gmünd, Germany*
Certomat®HK temperature-regulating device – *Sartorius, Göttingen, Germany*
Certomat®R shaking platform – *Sartorius, Göttingen, Germany*
CO_2-incubator STERI CULT 200 – *Forma Scientific Inc., San Bruno, USA*
Dounce homogenizer (1 ml) – *Kontes Glas Co., Vineland, USA*
DryGel Sr Slab Gel Dryer SE1160 – *Hoefer Scientific Instruments, San Francisco, USA*
Electrophoresis chamber SE 600 – *Hoefer Scientific Instruments, San Francisco, USA*
Eppendorf centrifuge 5417R – *Eppendorf AG, Hamburg, Germany*
Fluorometer Fusion – *Canberra-Packard, Dreieich, Germany*

Innova™ 4300 Incubator – *New Brunswick Scientific GmbH, Nürtingen, Germany*
Kinetic Microplate Reader VMax® – *Molecular Devices, Sunnyvale, USA*
Megafuge 1.0 – *Heraeus Sepatech, Langenselbold, Germany*
Mighty Small Dual gel caster SE 245 – *Hoefer Scientific Instruments, San Francisco, USA*
Mini vertical unit Mighty Small SE 250 – *Hoefer Scientific Instruments, San Francisco, USA*
PCR cycler T-Gradient – *Biometra, Göttingen, Germany*
PTC-200 Peltier Thermal Cycler – *Biozym, Hess.-Oldendorf, Germany*
Rocking platform – *Biometra, Göttingen, Germany*
Rolling platform TRM-V – *IDL, Nidderau, Germany*
Rotator GFL 3025 – *Gesellschaft für Labortechnik mbH, Burgwedel, Germany*
Thermomixer compact – *Eppendorf AG, Hamburg, Germany*
Titramax 100 – *Heidolph Instruments GmbH & Co.KG, Schwabach, Germany*
Transilluminator Biometra Ti1 – *Biometra, Göttingen, Germany*
UV-visible recording spectrometer UV-160 – *Shimadzu Deutschland GmbH, Duisburg, Germany*
Wide-Mini Sub® CELL GT – *Biometra, Göttingen, Germany*
Wobbling disk Polymax 1040 – *Heidolph Instruments GmbH & Co.KG, Schwabach, Germany*
Zeiss Axiovert 200 microscope – *Carl Zeiss AG, Oberkochen, Germany*

1.b Consumables

10 cm agar dishes – *Greiner Bio One, Frickenhausen, Germany*
15 ml tubes (bluecap) – *Greiner Bio One, Frickenhausen, Germany*
50 ml tubes (bluecap) – *Greiner Bio One, Frickenhausen, Germany*
96-well microplates, U-shaped bottom – *Sarstedt, Nümbrecht, Germany*
96-well Millipore plates (Millipore-MAHV N45) – *Millipore GmbH, Schwalbach, Germany*
384-well PCR plate – *Applied Biosystems, Darmstadt, Germany*
Amersham Hyperfilm™ ECL – *Amersham Biosciences, Freiburg, Germany*
BD Falcon™ 6 cm cell culture dishes – *Schuett24 GmbH, Göttingen, Germany*
BD Falcon™ 15 cm cell culture dishes – *Schuett24 GmbH, Göttingen, Germany*
Dialysis tubes – *GIBCO BRL, Karlsruhe, Germany*
Gloves – *Paul Hartmann AG, Heidenheim, Germany*

Luminometer tubes – *Sarstedt, Nümbrecht, Germany*
Microscope slides 76 x 26 mm – *Roth, Karlsruhe, Germany*
Nitrocellulose membrane Hybond™-ECL (0.45 µm) – *Amersham Biosciences, Freiburg, Germany*
Nunc™ Surface 6-well plates – *Nunc, Roskilde, Denmark*
Nunc™ Thermanox Plastic Coverslips, 25 mm diameter – *Nunc, Roskilde, Denmark*
Parafilm® M – *Brand GmbH & Co KG, Wertheim, Germany*
Quick Seal Tubes – *Beckmann GmbH, Munich, Germany*
Reaction tubes (1.5 mL, 2 mL) – *Eppendorf AG, Hamburg, Germany*
Safe-Lock PCR tubes (0.2 mL, 0.5 mL) – *Eppendorf AG, Hamburg, Germany*
Tips (10 µL) – *Eppendorf AG, Hamburg, Germany*
Tips (200 µL, 1 mL) – *Sarstedt, Nümbrecht, Germany*
Whatman paper – *Schleicher & Schüll, Dassel, Germany*

1.c Kits

Big Dye® Terminator v1.1 Cycle Sequencing Kit – *Applied Biosystems, Darmstadt, Germany*
Bradford Dye Reagent for Protein Assays – *Biorad, München, Germany*
EasyPure® DNA purification kit – *Biozym, Hess.-Oldendorf, Germany*
ECL Western Blotting Analysis System – *Amersham Biosciences, Freiburg, Germany*
Mini Quick Spin™ Oligo Columns – *Roche, Mannheim, Germany*
RNeasy Mini Kit – *QIAGEN, Hilden, Germany*
TaqMan® Gene Expression Master Mix – *Applied Biosystems, Darmstadt, Germany*
TNT T7 Coupled Reticulocyte Lysate System – *Promega, Mannheim, Germany*
Vectashield® Mounting Medium with DAPI – *Vector Laboratories, Burlingame, USA*

2. Chemicals

2.a Substances

25% glutaraldehyde – *Sigma Aldrich, Hamburg, Germany*
37% formaldehyde – *Applichem, Darmstadt, Germany*
87% glycerol – *Applichem, Darmstadt, Germany*
8-bromo-cAMP – *Sigma Aldrich, Hamburg, Germany*
[α-^{32}P]dCTP – *Hartmann Analytics, Braunschweig, Germany*
Acetic acid – *Applichem, Darmstadt, Germany*
Acrylamide – *Applichem, Darmstadt, Germany*
Adenosine triphosphate (ATP) – *Applichem, Darmstadt, Germany*
Agar – *GIBCO BRL, Karlsruhe, Germany*
Agarose (electrophoresis grade) – *Invitrogen, Karlsruhe, Germany*
Ammoniumpersulphate (APS) – *Applichem, Darmstadt, Germany*
Ampicillin – *Applichem, Darmstadt, Germany*
Aprotinin – *Applichem, Darmstadt, Germany*
Aqua ad injectabilia – *Braun, Melsungen, Germany*
Bis-acrylamide – *Applichem, Darmstadt, Germany*
β-Mercaptoethanol – *Applichem, Darmstadt, Germany*
Boric acid – *Applichem, Darmstadt, Germany*
Bovine serum albumin (BSA) – *Applichem, Darmstadt, Germany*
Bromophenol blue – *Sigma Aldrich, Hamburg, Germany*
Cesium chloride (CsCl) – *Applichem, Darmstadt, Germany*
Chloroform – *Applichem, Darmstadt, Germany*
Coomassie brilliant blue – *Sigma Aldrich, Hamburg, Germany*
Cyclosprin A – *Sigma Aldrich, Hamburg, Germany*
DEAE-Dextran – *Amersham Pharmacia, Uppsala, Sweden*
Deoxycholic acid – *Sigma Aldrich, Hamburg, Germany*
Digitonin – *Merck AG, Darmstadt, Germany*
Dimethyl sulfoxide (DMSO) – *Applichem, Darmstadt, Germany*
Deoxynucleoside triphosphates (dNTPs) – *Roche, Mannheim, Germany*
Di-potassium hydrogen phosphate (K_2HPO_4) – *Applichem, Darmstadt, Germany*
Di-sodium hydrogen phosphate (Na_2HPO_4) – *Applichem, Darmstadt, Germany*
Dithiothreitol (DTT) – *Applichem, Darmstadt, Germany*

D-Luciferin – *P.J.K., Kleinbittersdorf, Germany*
D-Saccharose – *Applichem, Darmstadt, Germany*
EDTA – *Applichem, Darmstadt, Germany*
EGTA – *Applichem, Darmstadt, Germany*
Ethanol – *Applichem, Darmstadt, Germany*
Ethidium bromide – *Applichem, Darmstadt, Germany*
Fetal calf serum – *GIBCO BRL, Karlsruhe, Germany*
Forskolin – *Sigma Aldrich, Hamburg, Germany*
GBX Fixation solution – *KODAK AG, Stuttgart, Germany*
Glucose – *Applichem, Darmstadt, Germany*
Glutathione agarose – *Sigma Aldrich, Hamburg, Germany*
Glycine – *Applichem, Darmstadt, Germany*
Glycogen – *Applichem, Darmstadt, Germany*
Glycylglycine – *Applichem, Darmstadt, Germany*
HEPES – *Applichem, Darmstadt, Germany*
Horse serum – *GIBCO BRL, Karlsruhe, Germany*
Hydrochloric acid (HCl) – *Applichem, Darmstadt, Germany*
Isoamylalcohol – *Applichem, Darmstadt, Germany*
Isopropanol – *Applichem, Darmstadt, Germany*
Isopropyl-β-D-thiogalactoside (IPTG)
L-[^{35}S]Methionine – *Hartmann Analytics, Braunschweig, Germany*
Leupeptin – *Applichem, Darmstadt, Germany*
Lithium chloride (LiCl) – *Sigma Aldrich, Hamburg, Germany*
L-Glutathione – *Sigma Aldrich, Hamburg, Germany*
Lysozyme – *Applichem, Darmstadt, Germany*
LX24 x-ray developer – *KODAK AG, Stuttgart, Germany*
Magnesium chloride (MgCl$_2$) – *Applichem, Darmstadt, Germany*
Magnesium sulphate (MgSO$_4$) – *Applichem, Darmstadt, Germany*
Metafectene – *Biontex, München, Germany*
Methanol – *Applichem, Darmstadt, Germany*
Nonidet-P40 – *Merck AG, Darmstadt, Germany*
Polyethylene glycol 6000 (PEG 6000) – *Applichem, Darmstadt, Germany*
Penicillin / Streptomycin – *GIBCO BRL, Karlsruhe, Germany*

Pepstatin A hemisulphate – *Applichem, Darmstadt, Germany*

Pepton from casein – *Applichem, Darmstadt, Germany*

Phenol (Tris saturated) – *Biomol, Hamburg, Germany*

Phenylmethylsulfonylfluorid (PMSF) – *Applichem, Darmstadt, Germany*

Poly(deoxyinosinic-deoxycytidylic) acid [poly(dI-dC)] – *Sigma Aldrich, Hamburg, Germany*

Ponceau S solution – *Applichem, Darmstadt, Germany*

Potassium chloride (KCl) – *Applichem, Darmstadt, Germany*

Potassium di-hydrogen phosphate (KH_2PO_4) – *Applichem, Darmstadt, Germany*

Protein G agarose – *Sigma Aldrich, Hamburg, Germany*

Sephadex G50 – *Amersham Biosciences, Freiburg, Germany*

Sepharose CL-4B – *Sigma Aldrich, Hamburg, Germany*

Silver nitrate ($AgNO_3$) – *Sigma Aldrich, Hamburg, Germany*

Skim milk – *Applichem, Darmstadt, Germany*

Sodium acetate trihydrate (NaAc) – *Applichem, Darmstadt, Germany*

Sodium borohydrate – *Applichem, Darmstadt, Germany*

Sodium carbonate (Na_2CO_3) – *Applichem, Darmstadt, Germany*

Sodium chloride (NaCl) – *Applichem, Darmstadt, Germany*

Sodium di-hydrogen phosphate (NaH_2PO_4) – *Applichem, Darmstadt, Germany*

Sodium dodecylsulphate (SDS) – *Applichem, Darmstadt, Germany*

Sodium hydroxide (NaOH) – *Applichem, Darmstadt, Germany*

Sodium thiosulphate ($Na_2S_2O_3$) – *Applichem, Darmstadt, Germany*

TEMED – *Applichem, Darmstadt, Germany*

Tris – *Applichem, Darmstadt, Germany*

Triton X100 – *Sigma Aldrich, Hamburg, Germany*

Trypsin / EDTA – *GIBCO BRL, Karlsruhe, Germany*

Tween 20 – *Applichem, Darmstadt, Germany*

Tween 80 – *Applichem, Darmstadt, Germany*

Xylene cyanol FF – *Sigma Aldrich, Hamburg, Germany*

Yeast extract – *Applichem, Darmstadt, Germany*

2.b Stock solutions and buffers

2.b.I. Stocks

All stock solutions were prepared in double-destilled H_2O if not stated differently.

8-bromo-cAMP	100 mM	250 mg in 5.81 mL H_2O
APS	10% (w/v)	1 g / 10 mL
ATP	200 mM	1 g in 9.075 mL H_2O
BSA	20 mg / mL	200 mg / 10 mL
Cyclosporin A	830 µM	1 mg in 0.1 mL 99% EtOH plus 20 µL Tween 80 drop in 1 mL RPMI
DTT	1 M	1.542 g / 10 mL
EDTA pH 8.0	0.5 M	46.53 g / 250 mL
EGTA pH 7.8 – 8.0	180 mM	3.423 g / 50 mL
Ethidium bromide	10 mg / mL	100 mg / 10 mL
Forskolin	10 mM	1 mg / 243.6 µL DMSO
HEPES	1 M	59.58 g / 250 mL
Glucose	0.5 M	9 g / 100 mL
Glycine	1 M	7.507 g / 100 mL
Glycylglycine pH 7.8	0.5 M	3.303 g / 50 mL
KCl	2 M	2.98 g / 20 mL
K_2HPO_4	0.5 M	4.35 g / 50 mL
K_2HPO_4	100 mM	8.709 g / 500 mL
KH_2PO_4	100 mM	3.402 g / 250 mL
LiCl	2 M	0.848 g / 10 mL
LiCl	4 M	1.696 g / 10 mL
$MgCl_2$	1 M	5.08 g / 25 mL
$MgSO_4$	1 M	12.324 g / 50 mL
NaAc pH 4.8	3 M	20.41 g / 50 mL
NaCl	5 M	73.05 g / 250 mL
Na_2HPO_4	75 mM	1.33 g / 100 mL
NaH_2PO_4	75 mM	1.03 g / 100 mL
NaOH	1 N	2 g / 50 mL
SDS	10% (w/v)	25 g / 250 mL

Tris-base	1 M	60.57 g / 500 mL

Stock solutions were stored at room temperature, except stocks of 8-bromo-cAMP, APS, ATP, BSA, DTT, and Forskolin. Aliquots were prepared of the latter ones and stored at -20°C. Cyclosporin A was kept at 4°C.

2.b.II. Antibiotics and protease inhibitors

Ampicillin (Amp)	5% (w/v)	0.5 g / 10 mL
Penicillin / Streptomycin	10,000 U/mL / 10,000 µg/mL – ready to use solution (GIBCO)	
Aprotinin	50 µg / µL	5 mg / 100 µL of 10 mM Tris/HCl pH8.0
Leupeptin	50 µg / µL	5 mg / 100 µL of 10 mM Tris/HCl pH8.0
Pepstatin	50 µg / µL	5 mg / 100 µL DMSO
PMSF	200 mM	348 mg / 10 mL 99% EtOH
PMSF	50 mM	125 µL 200 mM PMSF, 99% EtOH ad 500 µL

Stock solutions of antibiotics and protease inhibitors were stored at -20°C

2.b.III. General buffers

Stocks of Tris-base and HEPES were adjusted to different pH using hydrochloric acid (HCl).

PBS pH 7.4	1x	1 L
NaCl	140 mM	8.00 g
KCl	2.5 mM	0.20 g
Na_2HPO_4	8.1 mM	1.44 g
KH_2PO_4	1.5 mM	0.24 g

The buffer was autoclaved and stored at room temperature.

3. Biological material

3.a Bacteria strains

The chemically competent *Escherichia coli* strain DH5α was used for plasmid amplification and expression of recombinant GST-fusion proteins.

3.b Eurcaryotic cell lines

Hamster insulinoma tumor cells, clone HIT-T15, are an insulin-producing beta-cell line established by simian virus 40 transformation of pancreatic islet cells from Syrian hamster (Santerre et al., 1981). HIT-T15 cells are electrically excitable like neurons (Schwaninger et al., 1993a) and were used for transient transfection assays, immunocytochemistry experiments, and ChIP assays.

3.c Cell culture and culture media

3.c.I Buffers and solutions

LB-Amp medium		*100 mL (autoclaved)*
NaCl	1% (w/v)	1 g
Pepton	1% (w/v)	1 g
Yeast extract	0.5% (w/v)	0.5 g
Ampicillin	50 µg / mL	100 µL 5% stock solution

RPMI complete		*500 mL*
RPMI Medium		450 mL
fetal calf serum	10% (v/v)	50 mL
horse serum	5% (v/v)	25 mL
Penicillin / streptomycin	1% (v/v)	5 mL

The LB-Amp medium was stored at room temperature. The RPMI complete was stored at 4°C. Ampicillin was added freshly to the LB medium before use.

Materials and methods

3.c.II. Procaryotic cultures

Bacteria were cultured in LB-medium supplemented with ampicillin. To grow cultures of volumes up to 200 mL the Certomat®R shaking platform was used agitating at 180 rpm. The temperature was kept at 37°C using the Certomat®HK temperature-regulating device. Cultures with volumes higher than 200 mL were grown in the Innova™4300 Incubator agitating at 200 rpm.

LB-Amp agar plates were prepared by using LB medium containing 1.5% (w/v) agar. The solution was autoclaved for 20 min at 120°C. After cooling to 50°C the ampicillin (Amp) was added and the LB-Amp agar was poured into 10 cm dishes. The agar plates were hardened at room temperature and stored at 4°C.

3.c.III. Eucaryotic cultures

Adherent HIT-T15 cells were cultured in RPMI (Roswell Park Memorial Institute) complete medium. The cells were kept at 37°C in a CO_2-incubator with 95% (v/v) humidity and 5% (v/v) CO_2. 15-cm dishes (176.6 cm^2 surface) were used for the culture. The cells were splitted once a week at full confluence (~35x10^6 cells): Cell monolayers were washed once with phosphate-buffered saline (PBS), and incubated for 3-5 min with 3 mL trypsin / EDTA at 37°C. The reaction was stopped by addition of 7 mL RPMI complete medium. Cells were detached from the dish, centrifuged for 2 min at 310xg (Megafuge 1.0), and washed once with RPMI complete medium. The cells were seeded at a density of ~0.5 x 10^6 cells / cm^2. After three days the medium was renewed.

3.d Plasmids and oligonucleotides

3.d.I. Expression constructs

All expression constructs prepared and used in the present work are listed in table 1. Constructs kindly provided and generated by Ulrike Böer are shown in italics. All coding sequences for full lengths or fragments of human TORC1, TORC2, and TORC3 are based on the sequences provided kindly by Mark Labow (Novartis Pharmaceuticals, Suffern, NY, USA) and deposited in the GenBank database under GenBankAccession Numbers AY360171, AY360172, and AY360173, respectively. Expression plasmids for CREB wild-type or mutants are based on the coding sequence of Δ-CREB deposited in the GenBank database under GenBankAccession Number M27691.

Expression vectors for mammalian or bacterial expression were the following:
- pcDNA3.1   Invitrogen, Karlsruhe, Germany
- pSG424   (Sadowski and Ptashne, 1989)
- pGEX2T   GE Healthcare, Munich, Germany
- pHKnt   (Frampton et al., 1993)

The construct *TORC1* encodes for the full-length human TORC1 comprising 651 amino acids. The coding sequence was cloned into the mammalian expression vector pcDNA3.1 by use of the restriction sites *Bam*HI and *Xba*I. The construct FLAG-TORC1 encodes for the full-length human TORC1 containing a FLAG epitope (DYKDDDDK) between the first and the second amino acid. The FLAG epitope was inserted by use of a modified primer. The coding sequence was subcloned into pcDNA3.1 using restriction sites *Bam*HI and *Xba*I. The plasmids *GAL4-TORC1* and *GAL4-TORC1$_{1-44}$* encode the full-length human TORC1 and the first 44 amino acids of TORC1, respectively, fused C-terminally to the DNA-binding domain of the yeast transcription factor GAL4 (amino acids 1-147). The coding sequence of TORC1, either full-length or the first 44 amino acids, was subcloned into the mammalian expression vector pSG424 by use of the restriction sites *Bam*HI and *Xba*I. The expression constructs *TORC1$_{1-327}$* and *TORC1$_{\Delta 44}$* encode truncated forms of TORC1 with only the first 327 amino acids and without the first 44 amino acids, respectively. The coding sequence was subcloned into the expression vector pcDNA3.1 using the restriction sites *Bam*HI and *Xba*I.

The expression construct *TORC2* encodes the full-length human TORC2 comprising 694 amino acids, and the plasmid *TORC2$_{1-347}$* encodes for a truncated form of TORC2 with only the first 347 amino acids. The coding sequences for both were subcloned into the mammalian expression vector pcDNA3.1 by use of the restriction sites *Hind*III and *Xba*I. The plasmids *GAL4-TORC2* and *GAL4-TORC2$_{1-53}$* encode the full-length human TORC2 and the first 53 amino acids of TORC2, respectively, fused C-terminally to the DNA-binding domain of the yeast transcription factor GAL4 (amino acids 1-147). The coding sequence of TORC2, either full-length or the first 53 amino acids, was subcloned into the mammalian expression vector pSG424 by use of the restriction sites *Kpn*I and *Xba*I.

Materials and methods

The construct *TORC3* encodes the human full-length TORC3 protein comprising 620 amino acids and the plasmid *TORC3$_{1-310}$* encodes the truncated form of TORC3 comprising only the first 310 amino acids. Both coding sequences were subcloned into the mammalian expression vector pcDNA3.1 using the restriction sites *Kpn*I and *Xba*I. Expression plasmids *GAL4-TORC3* and *GAL4-TORC3$_{1-46}$* encode the full-length human TORC3 and the first 46 amino acids of TORC3, respectively, fused C-terminally to the DNA-binding domain of GAL4 (amino acids 1-147). The coding sequences was subcloned into the vector pSG424 using the restriction sites *Kpn*I and *Xba*I.

The construct *hamTORC1* encodes for the full-length TORC1 protein from Syrian hamster cloned from HIT-T15 cells in the present work. The coding sequence was subcloned into pcDNA3.1 using the restriction sites *Bam*HI and *Xba*I.

Three point mutations were introduced to Δ-CREB during the present work: the arginine residue at position 300 was substituted with alanine (R300A); the lysine residue at position 290 was substituted with glutamate (K290E) or alanine (K290A). The expression plasmids *CREB* and *CREB-R300A* encode the wild-type full-length human Δ-CREB and the full-length human Δ-CREB R300A mutant, respectively. The coding sequences comprising 327 amino acids were subcloned into pcDNA3.1 using the restriction sites *Bam*HI and *Xba*I. The expression constructs *GAL4-CREB*, *GAL4-CREB-R300A*, *GAL4-CREB-K290E*, and *GAL4-CREB-K290A* encode the full-length human Δ-CREB as wild-type or mutant (R300A, K290E, or K290A) fused C-terminally to the DNA-binding domain of GAL4 (amino acids 1-147). The coding sequence was subcloned into the expression vector pSG424 using the restriction sites *Bam*HI and *Sac*I.

The expression constructs *GST-CREB*, *GST-CREB-R300A*, *GST-CREB-K290E*, and *GST-CREB-K290A* were used to express recombinant GST-fusion proteins of CREB wild-type and mutants in *E.coli*. To allow the purification from *E.coli* a truncated form of CREB (wild-type and mutants) was used comprising amino acids 23 – 327. The coding sequences of human Δ-CREB wild-type and mutants (R300A, K290E, or K290A) were subcloned into the bacterial expression vector pGEX-2T using restriction sites *Bam*HI and *Eco*RI.

Amino acids 269-327 of human Δ-CREB are coding for the basic leucine zipper (bZip) structure. The constructs *GAL4-bZip*, *GAL4-bZip-R300A*, *GAL4-bZip-K290E*, and *GAL4-bZip-K290A* encode the bZip structure as wild-type or mutants (R300A, K290E, or K290A) fused C-terminally to the DNA-binding domain of the yeast transcription factor GAL4 (amino acids 1-147). The coding sequences were subcloned into the expression vector pSG424 using the restriction sites BamHI and XbaI.

The expression constructs VP16-bZip, VP16-bZip-R300A, VP16-bZip-K290E, and VP16-bZip-K290A encode the bZip structure of Δ-CREB as wild-type or mutants (R300A, K290E, or K290A) fused C-terminally to the transactivation domain of the viral protein VP16. The coding sequences were subcloned into the mammalian expression vector pHKnt using the restriction sites BamHI and EcoRI.

As internal control for luciferase reporter-gene assays the pGFPtpz-cmv® control vector (Caberra-Packard, Dreieich, Germany) was used. This expression vector codes for the green fluorescent protein (GFP) variant topaz under control of the cytomegalovirus promoter.

Table 1: Expression constructs.

The table lists all constructs used in the present work. Constructs prepared and kindly provided by Ulrike Böer are shown in italics. Information is given on the biological source of the coding sequence and the expression vector. Additional information on the expressed protein is given in the column *notes*.

name	source	vector	notes
TORC1	homo sapiens	pcDNA3.1	full length
FLAG-TORC1	homo sapiens	pcDNA3.1	FLAG-tag
GAL4-TORC1	*homo sapiens*	*pSG424*	*full length*
TORC1$_{1-327}$	homo sapiens	pcDNA3.1	aa 1-327
GAL4-TORC1$_{1-44}$	*homo sapiens*	*pSG424*	*aa 1-44*
TORC1$_{\Delta 44}$	homo sapiens	pcDNA3.1	aa 45-651
TORC2	homo sapiens	pcDNA3.1	full length
GAL4-TORC2	*homo sapiens*	*pSG424*	*full length*
TORC2$_{1-347}$	homo sapiens	pcDNA3.1	aa 1-347
GAL4-TORC2$_{1-53}$	homo sapiens	pSG424	aa 1-53
TORC3	homo sapiens	pcDNA3.1	full length
GAL4-TORC3	*homo sapiens*	*pSG424*	*full length*
TORC3$_{1-310}$	homo sapiens	pcDNA3.1	aa 1-310
GAL4-TORC3$_{1-46}$	homo sapiens	pSG424	aa 1-46
hamTORC1	cricetulus aureus	pcDNA3.1	cloned from HIT-T15
CREB	*homo sapiens*	*pcDNA3.1*	*aa 1-327*
CREB-R300A	*homo sapiens*	*pcDNA3.1*	*Arg300 --> Ala*
GAL4-CREB	*homo sapiens*	*pSG424*	*full length*
GAL4-CREB-R300A	*homo sapiens*	*pSG424*	*Arg300 --> Ala*
GAL4-CREB-K290E	homo sapiens	pSG424	Lys290 --> Glu
GAL4-CREB-K290A	homo sapiens	pSG424	Lys290 --> Ala
GST-CREB	*homo sapiens*	*pGEX-2T*	*aa 23-327*
GST-CREB-R300A	*homo sapiens*	*pGEX-2T*	*Arg300 --> Ala*
GST-CREB-K290E	homo sapiens	pGEX-2T	Lys290 --> Glu
GST-CREB-K290A	homo sapiens	pGEX-2T	Lys290 --> Ala
GAL4-bzip	*homo sapiens*	*pSG424*	*aa 269 - 327*
GAL4-bzip-R300A	*homo sapiens*	*pSG424*	*Arg300 --> Ala*
GAL4-bzip-K290E	homo sapiens	pSG424	Lys290 --> Glu
GAL4-bzip-K290A	homo sapiens	pSG424	Lys290 --> Ala
VP16-bZip	*homo sapiens*	*pHKnt*	*aa 269 - 327*
VP16-bZip-R300A	*homo sapiens*	*pHKnt*	*Arg300 --> Ala*
VP16-bZip-K290E	homo sapiens	pHKnt	Lys290 --> Glu
VP16-bZip-K290A	homo sapiens	pHKnt	Lys290 --> Ala

3.d.II. Luciferase reporter gene constructs

Luciferase reporter gene constructs used in this work are listed in table 2. The promoter sequences were cloned into the pXP2 vector carrying the promoter-less luciferase gene (Nordeen, 1988). *4xsomCRE-Luc* (Oetjen et al., 1994) consists of four repeats of the CRE-containing region -58 to -31 of the rat somatostatin gene in front of the truncated thymidine kinase promoter (-81 to +52) of the herpes simplex virus. *G5E1B-Luc* (Kruger et al., 1997) contains five repeats of an enhancer element, identified in yeast as binding site for the yeast transcription factor GAL4 (Webster et al., 1988), in front of the viral E1B TATA box (Lillie and Green, 1989; Liu and Green, 1990). *-711Fos-Luc* (Eckert et al., 1996) contains the promoter region -711 to +45 of the human *fos* gene (Konig et al., 1989). The CRE-site of *-711FosCREm-Luc* was mutated by site-directed mutagenesis. This mutation of the CRE-sequence prevented the binding of CREB (Sassone-Corsi et al., 1988). *NR4A2-Luc* (Conkright et al., 2003b) contains the promoter region -389 to +154 of the human *NR4A2* gene, whereas *NR4A2CREm-Luc* contains a destroyed CRE-site. Here, the plasmid was digested with *Aat*II cutting central in the CRE sequence and the 3'-protruding nucleotides were blunted using T4 DNA Polymerase (4.d). After religation the CRE sequence was destroyed. *BDNF4-Luc* and *BDNF4CREm-Luc* (Fang et al., 2003) were kindly provided by Dr. Hung Fang (National Research Council, Ottawa, Canada). These constructs contain the promoter region -242 to +306 of exon IV of the human *BDNF* gene (with reference to the nomenclature introduced by Pruunsild and colleagues (Pruunsild et al., 2007)).

Table 2: Luciferease reporter-gene constructs.
Promoter sequences to be used in luciferase reporter-gene assays were cloned into the vector pXP2. The biological source of the promoter and the respective region are listed.

name	source	region	
4xsomCRE-Luc	rattus norvegicus	oligomerized promoter region -58 to -31	
		28mer:	5'-GATCCTCCTTGGCTGACGTCAGAG AGAGAGTA-3'
G5E1B-Luc	s. cerevisiae	oligomerized enhancer element (GAL4-binding site)	
		17mer:	5'-CGGAGTACTGTCCTCCG-3'
-711Fos-Luc	homo sapiens	promoter region -711 to +45 of c-fos gene	
-711FosCREm-Luc	homo sapiens	promoter region -711 to +45 of c-fos gene (CRE mutated)	
NR4A2-Luc	homo sapiens	promoter region -389 to +154 of NR4A2 gene	
NR4A2CREm-Luc	homo sapiens	promoter region -389 to +154 of NR4A2 gene (CRE mutated)	
BDNF4-Luc	homo sapiens	promoter region -242 to +306 of Exon IV of BDNF gene	
BDNF4CREm-Luc	homo sapiens	promoter region -242 to +306 of Exon IV of BDNF gene (CRE muatated)	

3.d.III. Oligonucleotides

An overview of the combination of primers used for PCR cloning (4.c.II) of the constructs generated during the present work is presented in table 3. The combination with primers for site-directed mutagenesis (4.c.III) is indicated. The sequences of the primers for cloning and for site-directed mutagenesis are given in table 4 and table 5, respectively.

Table 3: Overview of the primer pairs used to generate the constructs in the present work.
The construct name is given in italics. The primer combinations for PCR cloning are presented. For constructs which were generated by site directed mutagenesis followed by PCR cloning all oligonucleotides used for this purpose are indicated.

name	PCR-cloning		site directed mutagenesis	
	primer forward	primer reverse	primer forward	primer reverse
TORC1	TORC1-BamH-f	TORC1-Xba-r		
FLAG-TORC1	FL-TORC1-BamH-f	TORC1-Xba-r		
TORC1$_{1-327}$	TORC1-BamH-f	TORC1_981_Xba-r		
TORC1$_{\Delta44}$	TORC1_45_BamH_f	TORC1-Xba-r		
TORC2	TORC2-Hind-f	TORC2-Xba-r		
TORC2$_{1-347}$	TORC2-Hind-f	TORC2_1041_Xba_r		
GAL4-TORC2$_{1-53}$	GAL4-T2-Kpn-f	GAL4-T2_1-53_Xba-r		
TORC3	TORC3-Kpn-f	TORC3-Xba-r		
TORC3$_{1-310}$	TORC3-Kpn-f	TORC3_930_Xba_r		
GAL4-TORC3$_{1-46}$	GAL4-T3-Kpn-f	GAL4-T3_1-46_Xba-r		
hamTORC1	TORC1-BamH-f	TORC1-Xba-r		
GAL4-CREB-K290E	CREB-Kpn-f	CREB-Sac-r	CREB-K290E-f	CREB-K290E-r
GAL4-CREB-K290A	CREB-Kpn-f	CREB-Sac-r	CREB-K290A-f	CREB-K290A-r
GST-CREB-K290E	GST-CREB-BamH-f	CREB-EcoR-r		
GST-CREB-K290A	GST-CREB-BamH-f	CREB-EcoR-r		
GAL4-bzip-K290E	GAL4-bZip-BamH-f	CREB327-Xba-r		
GAL4-bzip-K290A	GAL4-bZip-BamH-f	CREB327-Xba-r		
VP16-bZip-K290E	GST-bZip-BamH-f	CREB-EcoR-r		
VP16-bZip-K290A	GST-bZip-BamH-f	CREB-EcoR-r		
-711FosCREm-Luc	pXP2-Hind-f	pXP2-Xho-r	FosCREmut-f	FosCREmut-r

3.d.III.A Oligonucleotides used for PCR cloning

Synthetic oligonucleotides were purchased from Eurofins MWG Operon (Ebersberg, Germany). Table 4 lists the sequences in 5' – 3' direction of primers used for PCR cloning procedures (4.c.II).

Table 4: Synthetic oligonucleotides used for PCR cloning of expression constructs and luciferase reporter-gene constructs.

Sequences of oligonucleotides are presented in 5' – 3' direction. The direction (dir) of the primer with respect to the coding sequence is indicated: forward (for) and reverse (rev). The restriction sites are underlined and are indicated in the column *restr. site*. Stop-codons and Start-codons are shown in bold. Artificially inserted tags are shown in italics.

name	dir	sequence	restr. site
TORC1-BamH-f	for	GCGGGATCCCCACC**ATG**GCGACTTCGAACAATCCGC	BamHI
TORC1-Xba-r	rev	GCGTCTAGA**TTA**CAGGCGGTCCATCCGGAAGGT	XbaI
FL-TORC1-BamH-f	for	GCGGGATCCCCACC**ATG***GACTACAAGGACGACGACGACAAGG CGACTTCGAACAATCCGCGG*	BamHI
TORC1_981_Xba_r	rev	GCGTCTAGA**TTA**GACGCCAGCTGGGCGGTGCTG	XbaI
TORC2-Hind-f	for	GCGAAGCTTCCACC**ATG**GCGACGTCGGGGGCGAACGGG	HindIII
TORC2-Xba-r	rev	GCGTCTAGA**TTA**TTGGAGCCGGTCACTGCGGAA	XbaI
TORC1_45_BamH_f	for	GCGGGATCCCCACC**ATG**CTCCAGAAATCCCAGTACCTG	BamHI
TORC2_1041_Xba_r	rev	GCGTCTAGA**TTA**TAGGGAGGACTGCAGGGATGG	XbaI
GAL4-T2-Kpn-f	for	GAGGTACCCCATGGCGACGTCGGGGGCGAA	KpnI
GAL4-T2_1-53_Xba-r	rev	GATCTAGACTGTAACCGGGTGGAGCCGAT	XbaI
TORC3-Kpn-f	for	GCGGGTACCCCACC**ATG**GCCGCCTCGCCGGGCTCGGGC	KpnI
TORC3-Xba-r	rev	GCGTCTAGA**TTA**CAGTCTGTCAGCTCGAAACGT	XbaI
TORC3_930_Xba_r	rev	GCGTCTAGA**TTA**TGGGATGTTGTTCACACTATT	XbaI
GAL4-T3-Kpn-f	for	AAGGTACCCCATGGCCGCCTCGCCGGGCTC	KpnI
GAL4-T3_1-46_Xba-r	rev	GATCTAGATTGAACCCGCGACAGGGTGAG	XbaI
CREB-Kpn-f	for	ACAGCTGGCTAGCAATGGTACCG	KpnI
CREB-Sac-r	rev	TGATCATTACTTATCTAGAGCTC	SacI
GST-CREB-BamH-f	for	CGCGGATCCCAACAAATGACAGTTCAAG	BamHI
CREB-EcoR-r	rev	CGCGAATTC**TTATTA**ATCTGATTTGTGGCAG	EcoRI
GST-bZip-BamH-f	for	CGCGGATCCGCACGAAAGAGAGAGGTCC	BamHI
CREB327-Xba-r	rev	CGCTCTAGA**TTATTA**ATCTGATTTGTGGCAGTAAAGGTC	XbaI
GAL4-bZip-BamH-f	for	CGCGGATCCTAGCACGAAAGAGAGAGGTCC	BamHI
pXP2-Hind-f	for	TCCAAGCTCAGATCCAAGCTTGCG	HindIII
pXP2-Xho-r	rev	ATGCCAAGCTCAGATCTCGAGCGC	XhoI

3.d.III.B Oligonucleotides used for site-directed mutagenesis

Synthetic oligonucleotides used for site-directed mutagenesis by primerless PCR (4.c.III) were designed as complementary to each other. Sequences in 5' – 3' direction are listed in table 5. The primers CREB-K290E-f and CREB-K290E-r were used to substitute the lysine residue (K) at position 290 in the coding sequence of human Δ-CREB with glutamate (E): codon AAG (K) changed to GAG (E). The primers CREB-K290A-f and CREB-K290A-r were used to substitute the lysine residue (K) at position 290 in the coding sequence of human Δ-CREB with alanine (A): codon AAG (K) changed to GCG (A). The primers FosCREmut-f and FosCREmut-r were used to mutate the CREB-binding site (CRE) in the promoter of the human *fos* gene in the -711Fos-Luc construct (3.d.II). The original CRE sequence of the -711Fos promoter TGACGTTT was changed to TTAAACT.

Table 5: Synthetic oligonucleotides used for site-directed mutagenesis by primerless PCR.
The sequences of primers used for site-directed mutagenesis are presented in 5' – 3' direction. Modified nucleotides are highlighted in grey. Primers were designed as complementary to each other. The direction of the primer with respect to the coding sequence is indicated: forward (for) and reverse (rev).

name	direction	sequence
CREB-K290E-f	for	GTCGTAGAAAGGAGAAAGAATATGTG
CREB-K290E-r	rev	CACATATTCTTTCTCCTTTCTACGAC
CREB-K290A-f	for	GTCGTAGAAAGGCGAAAGAATATGTG
CREB-K290A-r	rev	CACATATTCTTTCGCCTTTCTACGAC
FosCREmut-f	for	GGTTGAGCCCGTTAAACTTACACTCATTCA
FosCREmut-r	rev	TGAATGAGTGTAAGTTTAACGGGCTCAACC

3.d.III.C Oligonucleotides used for sequencing

Primers designed for sequencing (4.e) of newly generated constructs are listed in table 6. The sequences are presented in 5' – 3' direction and the respective bases are indicated. The primer CMV-f anneals to the CMV promoter of the pcDNA3.1 expression vector and was used to analyze the sequence inserted downstream of the promoter. The primer pcDNA3-r anneals close to the multiple cloning site of the expression vector pcDNA3.1 and was used to analyze the sequence upstream of this site. The primers CRG1-f and CRG2-f anneal upstream of the coding sequence for the DNA-binding domain of GAL4 and upstream of the multiple cloning site, respectively, in the expression vector pSG424 and were used to analyze the in-frame insertion of subcloned sequences into this vector.

The primers TORC1-Seq2-f and TORC1-Seq4-f were used to verify the coding sequence for human TORC1. The primers hamTORC-Seq2-f, ham-TORC-Seq3-r, hamTORC-Seq4-f, and hamTORC-Seq5-r were used to analyze the sequence of the hamster TORC1 isoform cloned from HIT-T15 cells in the present work. To verify the promoter sequences inserted upstream of the luciferase gene the primer Luci-Seq-r was used.

Table 6: Synthetic oligonucleotides used to sequence newly generated constructs.
Sequences of synthetic primers are presented in 5' – 3' direction. The direction of the primer with respect to the coding sequence to be analyzed is indicated: forward (for) and reverse (rev). The bases to which the primers anneal are indicated in the column site of annealing.

name	dir	sequence	site of annealing
CMV-f	for	GAGGTCTATATAAGCAGAGC	pcDNA3.1 bases 798 - 817
pcDNA3-r	rev	AGGCACAGTCGAGGCTGATC	pcDNA3.1 bases 1050 - 1030
CRG1-f	for	ATTCCAGAAGTAGTGAGGAGGC	pSG424 bases 2830 - 2851
CRG2-f	for	ATCATCATCGGAAGAGAGTAG	pSG424 bases 3293 - 3313
TORC1-Seq2-f	for	ACCAATTCTGACTCCGCCCTG	TORC1 bases 492 - 513
TORC1-Seq4-f	for	CCCGTCCGCCTGCCCCCTGGT	TORC1 bases 1212 -1233
hamTORC-Seq2-f	for	CACAGGGGGCTCCCTCCCTGACCTCAG	hamTORC bases 723 - 749
hamTORC-Seq3-r	rev	GGGGTGGTGGCTGCTGTTGGGACA	hamTORC bases 1147 - 1124
hamTORC-Seq4-f	for	GCGGACACCAGCTGGAGGAGG	hamTORC bases 424 - 444
hamTORC-Seq5-r	rev	CCAGCTGGCGGGACAGCGCATTGG	hamTORC bases 1504 - 1481
Luci-Seq-r	rev	GGTTCCATCCTCTAGAGG	pXP2 bases 144-127

3.d.III.D Oligonucleotides used for quantitative real-time PCR

Primers used for quantitative real-time PCR in ChIP-assays as well as the TaqMan™ probes are listed in table 7. Primers CRE_ChIP_f and CRE_ChIP_r and the TaqMan™ probe CRE were designed to quantify amounts of 4xsomCRE-Luc plasmid (3.d.II) by quantitative real-time PCR (16.c). The primer CRE_ChIP_f anneals upstream of the inserted oligomerized CRE-promoter sequence to the bases 6071 – 6086 of the pXP2 vector. The primer CRE_ChIP_r anneals downstream of the oligomerized CRE-promoter sequence to the bases +1 – -14 of the thymidine kinase promoter. The TaqMan™ probe CRE anneals to bases -46 – -24 of thymidine kinase promoter. The primer GAL4_ChIP_r and the TaqMan™ probe GAL4 were designed to quantify amounts of G5E1B-Luc (3.d.II) by quantitative real-time PCR (16.c). G5E1B-Luc like the 4xsomCRE-Luc is based on the pXP2 vector, therefore the primer CRE_ChIP_f was used as well for the quantification of G5E1B-Luc. The primer GAL4_ChIP_r anneals downstream of the E1B TATA box to bases +58 of the pXP2 vector – +18 of the 20-bp spacer between the E1B TATA box and the pXP2 vector. The TaqMan™ probe GAL4 anneals to bases -2 of the E1B TATA box – +16 of the 20-bp spacer between the E1B TATA box and the pXP2 vector.

Table 7: Synthetic oligonucleotides and TaqMan™ probes for quantitative real-time PCR.
Sequences of synthetic primers and TaqMan™ probes are presented in 5' – 3' direction. The direction of the primer with respect to the coding sequence to be analyzed is indicated: forward (for) and reverse (rev). Labeling of the TaqMan™ probe with fluorophore and quencher is indicated.

name	direction	sequence
CRE_ChIP-f	for	GCAATAGCATCACAAATTTCACAAA
CRE_ChIP-r	rev	CCGCCCCGACTGCAT
GAL4_ChIP_r	rev	AATGCCAAGCTGGAATTCGA
TaqMan™ probe CRE	for	CGAATTCGCCGGATCTCGAGCTC modifications: 5'–Fluorescein; 3'–TAMRA
TaqMan™ probe GAL4	for	TAGAGGGTATATAATGATCCCCGGGTTACCGAG modifications: 5'–Fluorescein; 3'–TAMRA

3.d.III.E Oligonucleotides to generate probes

Synthetic oligonucleotides, that were used as double-stranded probes (11.c.I) for EMSAs, are listed in table 8. Sequences are presented in 5' – 3' direction.

Table 8: Synthetic oligonucleotides used as double-stranded probes for EMSAs.
Sequences of synthetic primers are presented in 5' – 3' direction. The direction of the primer with respect to the coding sequence to be analyzed is indicated: forward (for) and reverse (rev).

name	direction	sequence
SomCRE-f	for	5'-GATCCTCCTTGGCTGACGTCAGAGAGAGAGTA-3'
SomCRE-r	rev	5'-GATCTACTCTCTCTCTGACGTCAGCCAAGGAG-3'

3.e Enzymes and markers

3.e.I. Restriction endonucleases

Restriction enzymes used in the present work were purchased from Fermentas Life Sciences (St. Leon-Rot, Germany):

Enzyme	Activity	Recognition site
AatII	10 U / µL	5'-G A C G T^C-3'
BamHI	10 U / µL	5'-G^G A T C C-3'
Eco32I (EcoRV)	10 U / µL	5'-G A T^A T C-3'
EcoRI	10 U / µL	5'-G^A A T T C-3'
HindIII	10 U / µL	5'-A^A G C T T-3'
KpnI	10 U / µL	5'-G G T A C^C-5'
SacI	10 U / µL	5'-G A G C T^C-3'
XbaI	10 U / µL	5'-T^C T A G A-3'
XhoI	10 U / µL	5'-C^T C G A G-3'

All enzymes were used for restriction digest of double-stranded DNA in combination with the buffer Tango™ yellow (Fermentas Life Sciences, St. Leon-Rot, Germany) according to the instructions of the manufacturer.

3.e.II. Modifying enzymes

Enzyme	Activity	Supplier
Pfu DNA Polymerase	2.5 U / µL	Fermentas Life Sciences
T4 DNA Ligase	1 U / µL	Fermentas Life Sciences
T4 DNA Polymerase	3 U / µL	Fermentas Life Sciences
Klenow Fragment	2 U / µL	Fermentas Life Sciences
M-MLV Reverse Transcriptase	200 U / µL	Promega, Mannheim, Germany
AmpliTaq® DNA Polymerase	3 U / µL	Applied Biosystems, Darmstadt, Germany
RNase A	10 mg / µL	Fermentas Life Sciences
Proteinase K	10 µg / µL	Fermentas Life Sciences

Enzymes were used in combination with the respective reaction buffer (10x or 5x) provided by and according to the instructions of the manufacturer.

3.e.III. Markers

Markers and ladders were purchased from Fermentas Life Sciences (St. Leon-Rot, Germany).

DNA-ladders:
GeneRuler™ 100bp DNA Ladder
GeneRuler™ 100bp Plus DNA Ladder
GeneRuler™ 1kb DNA Ladder
Molecular weight markers:
PageRuler™ Prestained Protein Ladder

3.f Antibodies

Primary antibodies used for Western blots (6.d), immunocytochemistry (16.), and chromatin immunoprecipitation assays (17.):

name	source	clone	company / reference
panTORC (1-42)	rabbit	polyclonal	Calbiochem, Darmstadt, Germany
FLAG M2	mouse	monoclonal	Sigma Aldrich, Hamburg, Germany
VP16 (1-21)	mouse	monoclonal	Santa Cruz Biotechnology, Heidelberg, Germany
GAPDH (FL-335)	rabbit	polyclonal	Santa Cruz Biotechnology, Heidelberg, Germany
CREB-KID	rabbit	polyclonal	J.F. Habener, Boston, USA

The panTORC (1-42) antibody is directed against the first 42 amino acids of TORC. These amino acids are highly conserved among the three isoforms, therefore all TORCs are recognized by this antibody. The FLAG M2 antibody detects the proteins tagged with the FLAG epitope DYKDDDDK. The antibody VP16 (1-21) antibody is directed against amino acids 456 – 490 of the viral protein VP16. The antibody GAPDH (FL-335) was raised against full-length human glutaraldehyde-3-phosphate dehydrogenase (amino acids 1-335). The CREB-KID antibody is directed against the 60 amino acid stretch of CREB referred to as kinase inducible domain.

Materials and methods

Secondary antibodies used for immunolabeling in Western blots (6.d) and immunofluorescence (16.):

name	source	linked to	company
mouse IgG-HRP	sheep	horseradish peroxidase	Amersham Biosciences, Freiburg, Germany
rabbit IgG-HRP	donkey	horseradish peroxidase	Amersham Biosciences, Freiburg, Germany
rabbit IgG-Alexa®488	goat	AlexaFluor® 488	Invitrogen, Karlsruhe, Germany

The applications of these antibodies are presented in table 9. Sorted by application, the respective dilutions or amounts used are indicated.

Table 9: Applications and dilutions of primary and secondary antibodies.
The dilutions or amounts of the primary and secondary antibodies as used in the present work are given for the different application.

antibody	application	dilution / amounts
panTORC (1-42)	immunocytochemistry	1:1,500
CREB-KID	Western blot	1:25,000
GAPDH	Western blot	1:1,000
VP16 (1-21)	Western blot	1:200
FLAG M2	ChIP	10 µg
PanTORC (1-42)	EMSA supershift	1 µL, 3 µL, 6 µL
CREB-KID	EMSA supershift	1 µL, 3 µL, 6 µL
mouse IgG-HRP	Western Blot	1:10,000
rabbit IgG-HRP	Western Blot	1:10,000
Rabbit IgG-Alexa®488	immunocytochemistry	1:50

4. Working with DNA and RNA

Reactions were mostly performed with nuclease free H_2O. For that purpose Aqua ad injectabilia was used and is indicated in the following paragraphs by aqua.

4.a DNA gel electrophoresis

4.a.I. Buffers and solutions

TAE-buffer	1x	1 L
Tris	40 mM	4.84 g
EDTA	1 mM	2 mL of 0.5 M stock
Acetic acid	20 mM	1.14 mL

Loading buffer (blue)	6x	10 mL
Glycerol	30% (v/v)	3.39 mL of 87% glycerol
Bromophenolblue	0.25% (w/v)	0.025 g
Xylene Cyanol FF	0.25% (w/v)	0.025 g

All buffers and solutions were stored at room temperature.

4.a.II. Gel preparation and electrophoresis

Agarose gel electrophoresis was used to separate DNA fragments by an electric field. The percentage of agarose (in w/v) in the gel was chosen with respect to the size of fragments to be separated. 0.8% of agarose was chosen for samples ranging from 500 to 10,000 bp; 1.5% of agarose was chosen to separate DNA fragment ranging from 200 to 3,000 bp (Sambrook et al., 1989). The agarose was melted in TAE buffer and 0.5 µL ethidiumbromide solution per 10 mL of liquid gel were added. The gel was casted and hardened at room temperature. To electrophorese the DNA, the gel was placed in an electrophoresis chamber and covered with TAE buffer. Sufficient amount of 6x loading buffer was added to the sample and the mixture was placed into the slots. 5 µL of the desired ladder were loaded on the gel. Electrophoresis from cathode to anode was performed by applying a constant electric field of 80V (Biometra® Standard Power Pack P25). The DNA migrates through the gel and will be separated by size depending on the percentage of the agarose in the gel. The fluorescent dye ethidium bromide intercalates with the DNA thereby visualizing the DNA in the gel at UV-light of 366 nm wavelength. Analysis was perfomed using the BioDocAnalyze system including the transilluminator Biometra Ti1 and the belonging camera as well as the software BioDocAnalyze 2.0.

4.b DNA purification from agarose gels and solutions

To purify plasmid DNA from gels or out of solutions the Easy Pure® purification kit was used according to the manufacturer's instructions. Briefly, the desired bands marking the DNA were cut out of the gel. The agarose gel was melted at 55°C upon addition of 3 vol of a salt solution, in case that the DNA was in solution no heating was needed. Sufficient amount of a silica matrix was added to the suspension to bind the DNA. The matrix was pelleted and washed with an alcoholic solution. The pellet was dried and the DNA was eluted in 20 µL aqua. The silica matrix was removed by centrifugation and the eluted DNA was transferred to a new tube.

4.c PCR and site-directed mutagenesis

4.c.I. Buffer and solutions

dNTPs	10 mM	400 µL
dATP	10 mM	40 µL of 100 mM dATP
dGTP	10 mM	40 µL of 100 mM dGTP
dCTP	10 mM	40 µL of 100 mM dCTP
dTTP	10 mM	40 µL of 100 mM dTTP
Aqua		ad 400 µL

Aliquots of 10 µL were prepared and stored at -20°C.

4.c.II. Polymerase chain reaction (PCR)

Polymerase chain reaction (PCR) is used to amplify DNA *in vitro* by using specific primer pairs that define the sequence of interest. Initially, the double-stranded DNA is separated by heating to 95°C. The annealing temperature for the primers depends on their melting temperature and is ideally chosen 5°C lower. At 72°C the sequence is elongated from these primers by a thermostable DNA polymerase. This cycle is repeated up to 35 times. Ideally, the newly synthesized DNA is serving as template, too. Thereby, the DNA is amplified exponentially (Saiki et al., 1988). To assure high specificity and high yield of amplificate the *Pfu* DNA polymerase, which exhibits 3' → 5' proofreading activity, was used in the present work. PCR was performed using the gradient cycler T-Gradient. DNA templates were diluted to a concentration of 1 ng / µL using sterile H_2O, as determined photometrically (4.g).

Amounts of reagents per reaction were:

Template	1 ng
10x reaction buffer (with $MgCl_2$)	1x
DMSO	1.5 % (v/v)
Primer for	30 pmol
Primerrev	30 pmol
dNTPs	200 µM
Polymerase	1 U
Aqua	ad 50 µL

In general the program consisted of the following steps:

1. 3 min of initial denaturation	95° C
2. 1 min of denaturation	95° C
3. 1 min of annealing	depending on melting temperature of the primers
4. 2 min of elongation	72° C
5. 10 min of final elongation	72° C

Steps 2 to 4 were repeated 25 - 35 times.

The samples were electrophoresed on agarose gels (4.a). To proceed with cloning the desired DNA was purified from the gel (4.b).

4.c.III. Site-directed mutagenesis using primerless PCR

The primerless PCR was initially used to clone a gene from different single oligonucleotides. In this method fragments of the oligonucleotides can anneal to each other and function as primers. During cycles of denaturing, annealing and elongation the gene is built up gradually (Stemmer et al., 1995). This method was modified and used to introduce mutations at specific sites in the coding sequence of an expression plasmid or a promoter sequence in a reporter gene. The primer pairs were chosen such that during 2 independent PCRs 2 fragments were generated. The PCR products were electrophoresed

Materials and methods

on agarose-gel elctrophoresis (4.a) and the fragments were purified from the gel (4.b). The fragments each contain the mutation and overlap with each other at the site of mutation. 1 µL of these overlapping fragments were used in a primerless PCR. Since the two fragments are self annealing they function as primer and template. The whole construct is amplified after the primerless PCR by a PCR using the flanking primers (Figure 5).

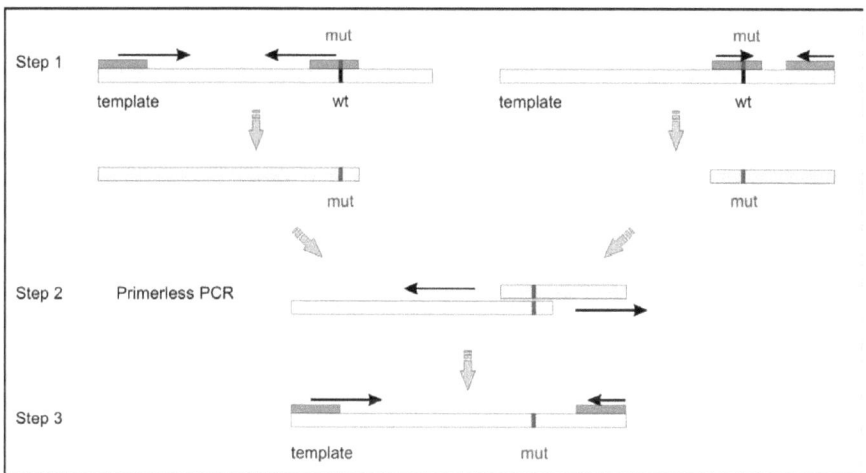

Figure 5: Schematic illustration of site-directed mutagenesis by primerless PCR.
For site-directed mutagenesis 3 different PCR steps were performed. In step 1 in two separate PCRs two fragments are generated, each containing the designated mutation. The mutation is introduced by a modified primer sequence. The generated fragments are overlapping at the site of mutation. In step 2 the two fragments are supplemented into a primerless PCR. The fragments are annealing to each other and serve as template and primer in one. Step 3 serves to amplify the complete fragment carrying the mutation.

4.d DNA modification

4.d.I. Restriction digest

Restriction endonucleases are used to cut DNA at specifically recognized nucleotide sequences. To digest plasmid DNA, 2 µg DNA were employed in the reaction. After a preparative PCR during cloning procedures the total amount of DNA resulting from PCR was used in restriction digests. The DNA was mixed with aqua and the respective amounts of 10x Tango™ yellow reaction buffer. This mixture was supplemented with 3 U of restriction endonuclease. The digest was carried out for 2 h in a waterbath at 37°C. To prepare samples for ligation, the digest was carried out overnight. The samples were analyzed by agarose gel electrophoresis.

4.d.II. Blunting

To blunt double stranded DNA with 3'-protruding termini the T4 DNA polymerase was used. This polymerase possesses 3' → 5' exonuclease activity. By omitting dNTPs in the reaction, the 5' → 3' synthesis is blocked. After restriction digest, resulting in 3'-protruding nucleotides, the DNA was purified from the gel and the total amount was used for the blunting reaction. Sufficient amounts of respective 5x reaction buffer were supplemented and 2.5 U of T4 DNA polymerase were added. The volume was filled up to 30 µL with aqua. The reaction was carried out for 20 min at 11°C and was stopped by incubation at 70°C for 10 min. Afterwards the DNA was purified from the solution and employed in a ligation reaction.

4.d.III. Ligation of DNA

The T4 DNA ligase is an enzyme that catalyzes the formation of a phosphodiester bond between juxtaposed 5'-phosphate and 3'-hydroxyl termini in double-stranded DNA. Ideally, fragments to be joined have fitting 3'- and 5'-protruding nucleotides as produced by various restriction endonucleases, though the enzyme also catalyzes ligation of blunt-ended DNA fragments. For the ligation, the insert was used in ~4-fold excess with respect to the vector. The fragments were mixed with aqua, sufficient amounts of respective 10x reaction buffer, and 1 U of T4 DNA Ligase. The reaction was performed in a waterbath at 16°C overnight. Afterwards, to amplify the ligated DNA, the total amount was used to transform chemically competent bacteria.

4.e Sequencing

DNA sequencing was performed to analyze the sequence of newly synthesized constructs. The DNA sequence was determined by the chain-termination method (Sanger et al., 1977) using the sequencing kit Big Dye® Terminator v1.1 Cycle Sequencing Kit. This kit contains fluorescently labeled didesoxynucleotides (ddATP, ddGTP, ddCTP und ddTTP) which lead to a termination of the DNA synthesis after incorporation. These ddNTPs are mixed with dNTPs in a relation that statistically allows the synthesis of fragments of all possible sizes. The PCR for sequencing was performed using the PTC-200 Peltier Thermal Cycler.

The following components were used for the reaction:
DNA	300 ng
Primer	10 pmol
BigDye® Mix	1.8 µL
DMSO	0.5 µL
Aqua	ad 10 µL

The PCR program used for the sequencing contained the following steps:
1.	2 min of initial denaturation	94° C
2.	15 sec of denaturation	96° C
3.	15 sec of annealing	56.5° C
4.	4 min of elongation	60° C
5.	7 min of final elongation	72° C

Steps 2 – 4 were repeated 24 times.

Subsequently, the PCR samples were purified by gel filtration chromatography. The samples were filled with aqua to a final volume of 40 µL. The Sephadex (Sephadex G50, Amersham Biosciences) was preswollen in a 96-well Millipore plate (Millipore-MAHV N45) and washed once with H_2O. The samples were added on top of the sephadex and were purified by centrifugation for 5 min at 150xg (Beckmann GS-6). Afterwards the samples were analyzed by capillary electrophoresis with respect to their size and the nucleotides were determined by fluorescence analysis using the ABI PRISM 3100 Genetic Analyzer.

4.f RNA extraction and reverse transcription

Total RNA was extracted from HIT-T15 cells using the RNeasy Mini Kit according to the manufacturer's instructions. HIT-T15 cells were cultured in 6-cm dishes (28.27cm² surface) and seeded at a density of ~0.2 x 10^6 cells per cm². 48h after plating the medium was removed and the cells were washed twice with PBS. The cells were lyzed immediately by scraping the cells in buffer RLT and transferring them to a 1.5 mL tube. After centrifugation at 20,800x*g* (Eppendorf 5417R) at 4°C, the supernatant was transferred to a new tube and mixed with 1 vol of 70% ethanol. The suspension was poured into a RNeasy spin column to bind the DNA and RNA from the suspension to the membrane of the column. The suspension was passed through the column by centrifugation. Subsequently, the DNA was digested directly on the column using the RNase-free DNase set. Digested DNA and the enzyme were removed from the membrane by washing with buffer RW1. Afterwards the membrane was washed twice with buffer RPE and the RNA was eluted from the membrane using aqua. The concentration of RNA was determined by UV-absorption (4.g). cDNA was synthesized from the total RNA by reverse transcription using oligo(dT)$_{15}$ primer and M-MLV Reverse Transcriptase, according to the instructions of the manufacturer. 2 µg RNA were mixed with 1 µg oligo(dT)$_{15}$ primer. The volume was filled up to 10 µL with aqua. At 70°C the single-stranded RNA was denatured for 5 min. Subsequently, the samples were put on ice.

The following components were added to the RNA – primer mixture to generate cDNA:

RNA-primer mixture	10 µL
M-MLV 5x reaction buffer	5 µL
Recombinant RNasin® Ribonuclease Inhibitor	25 U
M-MLV Reverse Transcriptase	200 U
Aqua	ad 25 µL

The reaction was performed for 1 h at 42°C.

The generated cDNA was used for analysis of the mRNA levels of the TORC proteins in HIT-T15 cells by PCR using primers for human TORC1, TORC2, and TORC3. In detail: TORC1-BamH-f and TORC1-Xba-r for TORC1, TORC2-Hind-f and TORC2-Xba-r for TORC2, TORC3-Kpn-f and TORC3-Xba-r for TORC3 (for sequence details see table 4). Amounts of TORC amplificates were normalized to β-actin levels amplified by the oligonucleotides ham-β-actin-f 5'-GATATCGCTGCGCTCGTTGTC-3' and ham-β-actin-r 5'-CCTCAGGGCAACGGAACC-3'. The PCR results were analyzed by agarose gel electrophoresis followed by densitometric analysis of the bands.

Materials and methods

4.g Quantification of DNA and RNA

The concentration and purity of DNA and RNA was determined by UV-absorption at 260 nm using a UV-visible recording spectrometer (Shimadzu UV-160). The measurement was performed in quartz cuvettes in a total volume of 1 mL TE-buffer (5.c) or Aqua.
The following calculations were used:

$$c\ (\mu g\ /\ \mu L) = OD_{260}\ x\ 50\ x\ dilution\ factor\ \text{for DNA, and}$$
$$c\ (\mu g\ /\ \mu L) = OD_{260}\ x\ 40\ x\ dilution\ factor\ \text{for RNA.}$$

The ratio between absorption at 260 nm and 280 nm gives a measure of the purity of the plasmid DNA. The ratio should have range of 1.8 to 2.0.

5. Amplification of plasmid DNA

5.a Transformation of chemically competent E.coli

The bacteria were stored at -80°C. For transformation the bacteria were thawed on ice. After ligation the total amount of DNA or, if the concentration was known, 100 ng of plasmid DNA were added to 30 µL of bacteria and kept for 30 min on ice. The heat shock was performed for 30 sec at 42°C. Afterwards the mixture was kept for 2 min on ice. 400 µL LB without antibiotics were added to the bacteria and the suspension was incubated for 1 h at 37°C, agitating at 400 rpm (Thermomixer compact). 200 µL of the bacteria were plated on a LB-Amp agar plate and incubated at 37°C in an incubator.

5.b Mini preparation

5.b.I. Buffers and solutions

Buffer M1		100 mL
Tris/HCl pH 8.0	25 mM	2.5 mL of 1 M stock
Glucose	50 mM	10 mL of 0.5 M stock
EDTA pH 8.0	10 mM	2 mL of 0.5 M stock
Lysozyme		2 mg / 1 mL buffer M1

Buffer M2		5 mL
NaOH	0.2 N	1 mL of 1 N stock
SDS	1% (w/v)	500 µL of 10% SDS
H$_2$O		3.5 mL

Buffer M3		20 mL
NaAc pH 4.8	3 M	8.165 g / 20 mL

All buffers were stored at room temperature. Lysozyme was added freshly to buffer M1 before use.

5.b.II. Procedure

To control successful cloning and mutagenesis procedures, DNA preparation in small scale range was performed by alkaline lysis and alcoholic precipitation (Sambrook et al., 1989). Single colonies were transferred from LB-Amp agar plates to 3 mL LB-Amp and grew overnight at 37°C. 1.5 mL of the culture were transferred to an reaction tube on ice. The bacteria were harvested by centrifugation at 3,200xg (Eppendorf 5417R) at 4°C. The supernatant was discarded and the pellet was resuspended in 150 µL buffer M1. The suspension was kept for 10 min on ice to lyse the cells. 200 µL of buffer M2 were added quickly and after 5 min the suspension was neutralized by slow addition of buffer M3. After further 5 min on ice the cell debris was pelleted by centrifugation for 10 min at 20,800xg (Eppendorf 5417R) at 4°C. The supernatant was transferred to a new reaction tube and the DNA was separated from proteins by addition of 200 µL phenol and 200 µL chloroform / isoamylalcohol (24:1). The mixture was vortexed for 30 sec. After 5 min centrifugation at 20,800xg (Eppendorf 5417R) at 4°C, the aqueous phase containing the DNA was transferred to a new reaction tube. 1/10 vol of buffer M3 and 2 vol 99% ethanol were added and the DNA was precipitated for 20 min at -20°C. The samples were centrifuged for 30 min at 20,800xg (Eppendorf 5417R) at 4°C. The pellet was washed with 70% ethanol and centrifuged again for 10 min at 20,800xg (Eppendorf 5417R) at 4°C. Afterwards the pellet was dried and resolved in 21 µL RNaseA (0.1 mg / mL).

5.c Maxi preparation

5.c.I. Buffers and solutions

STE buffer		100 mL
Tris/HCl pH 8.0	50 mM	5 mL of 1 M stock
Saccharose	25% (w/v)	25 g
EDTA pH 8.0	1 mM	0.2 mL of 0.5 M stock

Triton-mix		100 mL
Tris/HCl pH 8.0	50 mM	5 mL of 1 M stock
Triton X100	0.1% (v/v)	0.1 mL
EDTA pH 8.0	60 mM	12 mL of 0.5 M stock

PEG solution		100 mL
PEG 6000		30 g
NaCl	1.5 M	30 mL of 5 M stock

TNE buffer		100 mL
Tris/HCl pH 8.0	10 mM	1 mL of 1 M stock
NaCl	10 mM	0.2 mL of 5 M stock
EDTA pH 8.0	1 mM	0.2 mL of 0.5 M stock

TE buffer		1 L
Tris/HCl pH 8.0	10 mM	10 mL of 1 M stock
EDTA pH 8.0	1 mM	2 mL of 0.5 M stock

All buffers and solutions were stored at 4°C.

5.c.II. Procedure

The method was performed as described by Sambrook et al. (1989). 1L overnight culture of transformed bacteria was used to amplify large scales of plasmid DNA. The bacteria were harvested by centrifugation for 15 min at 2,500xg at 4°C, using the Beckmann J2HS centrifuge (rotor JA14) with 250 mL tubes. The supernatant was discarded and each pellet was resuspended in 11.25 mL STE buffer. Two pellets were pooled and 1.5 mL lysozyme solution (60 mg/mL STE buffer) were added. The tubes were kept 20 min on ice to lyse the cells. 1.8 mL 0.5 M EDTA were added per tube. After 5 min on ice, the tubes were filled up with 14.4 mL Triton-mix and were kept again 30 min on ice. After 1 h centrifugation at 31,000xg at 4°C (Beckmann J2HS centrifuge [rotor JA20]) the supernatant was pooled in one 250 mL tube, mixed with 40 mL PEG solution to precipitate the plasmid DNA and kept again 1 h on ice. After centrifugation for 10 min at 15,300xg at 4°C (Beckmann J2HS centrifuge [rotor JA14]), the supernatant was discarded and the pellet was dried. This pellet was resolved in 10.5 mL TNE buffer. 10.9 g CsCl and 150 µL ethidium bromide (10 mg/mL) were added per sample and the resulting suspension was transferred to a Quick Seal Tube. The tube was heat-sealed (Beckmann Tube Sealer) and the sample was ultracentrifuged for 20 h at 370,000xg at room temperature (Beckmann L8-70M Ultracentrifuge [rotor 70.1Ti]). Due to the density formed by CsCl, the plasmid DNA will form a distinct band during ultracentrifugation. This band was afterwards collected in a 15 mL bluecap using a syringe. To remove the ethidium bromide, the sample was washed 4-5 times with 1 vol of isoamylalcohol. The isoamylalcohol was added to the DNA-containing solution, the sample was mixed and centrifuged briefly at 2,500xg (Megafuge 1.0). The upper alcoholic phase containing the ethidium bromide was removed. To remove the CsCl the DNA was dialysed 2 times 12 h against TE-buffer (1 L per sample) at 4°C. The concentration, purity, and correctness of the plasmid DNA were determined by UV-absorption and restriction digest.

6. Analysis of proteins

6.a SDS-PAGE

6.a.I. Buffers and solutions

Stacking gel buffer		200 mL
Tris/HCl pH 6.8	0.5 M	12.144 g
SDS	0.4% (w/v)	8 mL of 10% SDS

Separating gel buffer		200 mL
Tris/HCl pH 8.8	1.5 M	36.342 g
SDS	0.4% (w/v)	8 mL of 10% SDS

Acrylamide solution		250 mL
Acrylamide	29.2% (w/v)	73 g
Bis-acrylamide	0.8% (w/v)	2 g

Tank buffer		1L
Tris	25 mM	3.03 g
Glycine	192 mM	14.41 g
SDS	0.1% (w/v)	10 mL of 10% SDS

Laemmli-loading buffer	2x	100 mL
Tris/HCl pH 6.8	160 mM	16 mL of 1 M stock
SDS	4 % (w/v)	4 g
Glycerol	10 % (v/v)	11.4 mL 87% Glycerol
Bromophenolblue	0.05 % (w/v)	50 mg
β-Mercaptoethanol	10 % (v/v)	10 mL

All buffers were stored at room temperature, except the Laemmli-loading buffer of which aliquots were prepared and stored at -20°C. The acrylamide solution was stored at 4°C.

6.a.II. Gel preparation and electrophoresis of proteins

Discontinuous sodiumdodecylsulphate polyacrylamide gel electrophoresis (SDS-PAGE) was initially described in 1970 (Laemmli, 1970). The method was here performed with some modifications according to Sambrook et al. (1989).

To cast the gels (8 cm x 7 cm) the Mighty Small Dual gel caster SE 245 was used together with the belonging alumina and glass plates as well as spacers and combs. Plates, spacers and the comb were rinsed with 70% ethanol to remove any contamination and remnants of fat. The plates separated by the spacers were fixed in the clamp assembly. By inserting it into the casting cradle the gasket seals the under edge of the plates. The separation gel was prepared and poured between the plates. To smooth the upper edge of the separation gel 99% ethanol was pipetted on the top. After polymerization of the separation gel the ethanol was removed and the remaining space between the glass plates was rinsed with H_2O. The stacking gel was prepared and poured on top of the separation gel. The comb forming the slots for the samples was inserted immediately.

Composition of stacking gel and separation gel:

stacking gel	4%	separation gel	10%
H_2O	3.13 mL	H_2O	4.2 mL
stacking gel buffer	1.25 mL	separation gel buffer	2.5 mL
acrylamide solution	620 µL	acrylamide solution	3.3 mL
APS	50 µL	APS	100 µL
TEMED	7.5 µL	TEMED	15 µL

The percentage of acrylamide supplemented to the gel depended on the molecular weights of the proteins to be analyzed by SDS-PAGE. Proteins of higher molecular weight (50 – 130 kDa) were electrophoresed on gels containing 8% acrylamide. Proteins in the range of 30 – 100 kDa were analyzed on gels containing 10% acrylamide, and proteins of lower molecular weight (20 – 70 kDa) were electrophoresed on 12% gels.

For the electrophoresis the mini vertical unit Mighty Small SE 250 was used. The gel was fixed with clamps to the chamber. The upper chamber was filled completely with tank buffer and the lower buffer chamber was filled to ¾. The comb was removed carefully and the slots were washed with tank buffer using a syringe. Each 1 vol of 2x Laemmli loading buffer was added to the samples and these were denatured for 10 min at 99°C, 1,000 rpm

(Eppendorf thermomixer). Afterwards the samples were briefly centrifuged and filled into the slots. The lid was put on the unit and connected to the power supply. A constant current of 15 mA per gel was applied until the samples reached the border between stacking gel and separation gel. Now the current was set to 25 mA per gel until the blue running front, marking the lower border of proteins in the electric field, reached the lower edge of the gel. The unit was disassembled; the gel was carefully removed from the plates and transferred to a dish for further treatment.

6.b Staining of proteins with Coomassie

6.b.I. Buffers and solutions

Stain solution		1 L
Coomassie brilliant blue	0.25 % (w/v)	2.5 g
Methanol	40 % (v/v)	400 mL
Acetic acid	10 % (v/v)	100 mL

Destain solution		1 L
Isopropanol	25 % (v/v)	250 mL
Acetic acid	10 % (v/v)	100 mL
Methanol	10 % (v/v)	100 mL

All solutions were stored at room temperature.

6.b.II. Procedure

Coomassie staining allows the detection of proteins in polyacrylamide gels (detection limit 100 -1,000 ng per band). The proteins were electrophoresed by SDS-PAGE. The gel was incubated for 20 min in stain solution under agitation on a rocking platform at room temperature. To destain the gel and thereby visualize the Coomassie-stained proteins the gel was incubated in destain solution for several hours with agitation on a rocking platform at room temperature. The destain solution was changed frequently. To conserve the gel, it was equilibrated to H_2O and dried in cellophane.

6.c Staining of proteins with silver

6.c.I. Buffers and solutions

Fixation solution		100 mL
Ethanol	30% (v/v)	30 mL
Acetic acid	10% (v/v)	10 mL

Sensitization solution		100 mL
Ethanol	30% (v/v)	30 mL
NaAc	0.5 M	6.804 g
Formaldehyde	0.5% (v/v)	1.33 mL of 37% formaldehyde
$Na_2S_2O_3$	0.2% (w/v)	0.2 g

Staining solution		100 mL
$AgNO_3$	0.1% (w/v)	0.1 g
Formaldehyde	0.02% (v/v)	54 µL of 37% formaldehyde

Developer		100 mL
Na_2CO_3	2.5% (w/v)	2.5 g
Formaldehyde	0.01% (v/v)	27 µL of 37% formaldehyde

Stop solution		100 mL
Acetic acid	5% (v/v)	5 mL

All solutions were stored at room temperature. The staining solution was kept in the dark.

6.c.II. Procedure

Staining of proteins using silver nitrate is a very sensitive method to visualize very low amounts of proteins (detection limit 1 - 10 ng per band) in polyacrylamide gels. The method was adapted from the silver staining previously described (Heukeshoven and Dernick, 1988) and was performed as described in the dissertation of Berit Brüstle, University of Konstanz.

Proteins were electrophoresed by SDS-PAGE. The gel was incubated in fixation solution to wash out SDS completely from the gel. The incubation was performed for 3 h agitating at room temperature (rocking platform). Afterwards the gel was sensitized for 30 min using sodium thiosulphate in the sensitization solution to maintain a reductive state of the gel during the staining procedure. The gel was washed 3 times for 10 min with H_2O to remove excess thiosulfate. The staining was performed for 1 h at room temperature using silver nitrate. Silver ions bind to proteins under alkaline conditions; these are reduced to silver during the development. To develop the gel, sodium bicarbonate and formaldehyde are used. As soon as the bands are visible the staining process is stopped by a strong change in pH using acetic acid. To conserve the gel, it was equilibrated to H_2O and dried in cellophane.

6.d Western blot

6.d.I. Buffers and solutions

Buffer A (anode buffer)		1 L
Tris, pH 11.3	300 mM	36.3 g
Methanol	20% (v/v)	200 mL

Buffer B (blot buffer)		1 L
Tris, pH 10.5	25 mM	3.03 g
Methanol	20% (v/v)	200 mL

Buffer C (cathode buffer)		1 L
Tris, pH 9.0	25 mM	3.03 g
Methanol	20% (v/v)	200 mL

PBS-T		1 L
Tween	0.1% (v/v)	1 mL
PBS		ad 1 L

Blocking solution		50 mL
Skim milk powder	5% (w/v)	2.5 g
PBS-T		ad 50 mL

Antibody solution		10 mL
Skim milk powder	1% (w/v)	0.1 g
PBS-T		ad 10 mL

All buffers were stored at room temperature. The blocking solution and the antibody solution were prepared freshly prior to use. The pH of buffer C was adjusted to 9.0 using boric acid.

6.d.II. Blotting

To analyse proteins by Western blot the samples were electrophoresed by SDS-PAGE (6.a). The gel was equilibrated to buffer C, agitating for 20 min at room temperature (rocking platform). The nitrocellulose membrane (Hybond™-ECL Nitrocellulose membrane [0.45 µm]) was equilibrated to buffer B for 10 min. Whatman paper were soaked with buffer A, buffer B, and buffer C.

To perform the semi-dry electrophoretic transfer of the proteins from the gel to the nitrocellulose membrane the paper, the gel and the membrane were arranged as follows in the blot chamber:

<u>cathode -- negative pole</u>
4 papers soaked with buffer C
Gel
Membrane
2 papers soaked with buffer B
2 papers soaked with buffer A
<u>anode -- positive pole</u>

The blot was performed by application of 1.2 mA per cm² membrane. Duration of the electrophoresis depended on the size of the proteins and varied between 45 min and 1 h. Afterwards the membrane was treated with Ponceau S solution to check for successful protein transfer. Ponceau S binds reversibly to the positively charged amino acids of the proteins on the membrane. The membrane was destained using PBS-T. Washing was performed 3 times for 10 min, agitating at room temperature on a rocking platform.

6.d.III. Immunodetection

To fill the unoccupied binding sites on the nitrocellulose, the membrane was treated for 1 h with blocking solution, agitating at room temperature on a rocking platform. Subsequently, the membrane was washed briefly with PBS-T. The primary antibody was diluted as desired in antibody solution (3.f; table 9) and the membrane was incubated overnight, agitating at 4°C. The next day the membrane was washed 3 times for 10 min with PBS-T, agitating at room temperature, to remove unbound primary antibody. The secondary antibody (3.f; table 9), as supplied with the ECL Western Blotting Analysis System (Amersham Biosciences), was diluted 1:10,000 in antibody solution. The membrane was incubated for 1 h, agitating at room temperature on a rocking platform. To remove unbound antibody the membrane was washed again 3 times for 10 min with PBS-T. The stained proteins were visualized by enhanced chemiluminescence using ultra-sensitive films (Amersham Hyperfilm™ ECL) and the ECL Western Blotting Detection Reagent (Amersham Biosciences) as recommended by the manufacturer. The solutions A and B were mixed to equal volumes and 1 mL of the mixture was used per nitrocellulose membrane (6 cm x 9 cm). The membrane was incubated for 5 min. Excess of the solution was removed carefully from the nitrocellulose; the membrane was wrapped in plastic film and placed in a photo cassette. Films were exposed to the membrane for varying periods depending on the reactivity of the antibody and the amount of protein. The photo films were developed for 1 min by rinsing with LX24 X-ray developer, washed for 2 min in H_2O, and fixed for 2 min in GBX fixation solution.

6.f Analysis of radioactively labeled proteins and probes

The radioactively labeled proteins were electrophoresed by SDS-PAGE, whereas the probes used in EMSAs were electrophoresed on non-denaturing gels (11.b – 11.d). In each case, the gel was dried using a gel dryer (DryGel Sr Slab Gel Dryer SE1160). Therefore the gel was placed on a Whatman gel-blotting paper and covered with plastic film. The humidity was removed by vacuum. The dried gels were exposed to a phosphor-imager screen (BAS-MS 2325, FUJIFILM). The exposure time varied between 3 h and 2 days depending on the incorporated isotope. The phosphor-imager screen was read-out using the phosphor-imager device BAS-1800II (FUJIFILM) and the software AIDA Version 4.15.025. For the distinct analysis of weaker signals the dried gels were exposed to X-ray films for 10 to 30 days. The films were developed as described in the section Western blot (6.d.III).

6.g Quantification of proteins

6.g.I. Bradford assay

The concentration of proteins in solution was determined using the Dye Reagent For Protein Assays (Biorad). This reagent contains the dye Coomassie Brilliant Blue G-250 which exists in two forms of different color, red and blue. The red one is changing to the blue form upon binding to protein, which allows the photometric measurement of protein concentration (Bradford, 1976). In the present work, the concentration of proteins purified by affinity chromatography and eluted from the agarose (7.b – 7.c) was determined by a Bradford assay. The Dye Reagent was diluted 1:5 in H_2O. BSA was used to generate a calibration curve ranging from 20 µg to 0.3125 µg; therefore, dilution series in 7 steps of 1:2 was prepared of BSA starting with 1 µg / µL in H_2O. The samples were diluted in 2 steps of 1:4 for the measurement. The assay was performed on 96-well microplates. Each sample as well as the calibration curve was measured in triplicate. 20 µL of sample or standard was added per well and 180 µL of the diluted Dye Reagent were supplemented. The protein concentration was determined by photometric measurement at 550 nm using the Kinetic Microplate Reader VMax® (Molecular Devices). To analyze the raw data the software SoftMax (Molecular Devices) was used.

6.g.II. Semi-quantitative SDS-PAGE

Different amounts of purified proteins in solution (1µL – 10 µL) and defined amounts of BSA (125 – 1,000 ng) were denatured in sufficient amount of 2x Laemmli loading buffer. The samples were electrophoresed by SDS-PAGE (6.a) and the gel was stained with Coomassie (6.b).

6.g.III. Quantification of proteins in small volumes

To quantify the protein concentration of cell extracts with volumes smaller than 50 µL, a different procedure was applied. Nuclear and cytosolic extracts from HIT-T15 cells were prepared as described in section 15. To determine the concentration of 1 µL cell extract the absorbance at 595 nm was measured photometrically in a 1 ml quartz cuvette using a UV-visible recording spectrometer (Shimadzu UV-160). Here 1 µL of cell extract was mixed with 1 mL of Dye Reagent For Protein Assays (Biorad) diluted 1:5 in H_2O. Beforehand a dilution series of BSA ranging from 50 µg to 1.5625 µg was measured respectively and the values were used to create a calibration curve. This calibration curve was applied afterwards to read out the concentration of the protein measurement.

7. Purification of GST-fusion proteins

In the present work, isolated GST-fusion proteins of CREB were used to analyze effects of lithium and magnesium on the interaction between CREB and TORC in a cell-free condition in GST pull-down assays (9.b). Additionally, GST-fusion proteins of CREB were employed in electrophoretic mobility shift assays (11.d) to investigate the DNA-binding of CREB. The first 44 amino acids of TORC1 were expressed as GST-fusion protein to investigate the effects of lithium on the tetramerization of TORC1 in crosslinking assays (10.b.I).

7.a Screening

The cDNA of the protein of interest was subcloned into the vector pGEX2T resulting in expression vectors coding for the protein of interest fused to glutathione S-transferase (GST). The coding sequence for the GST-fusion protein is under control of the *lac* operon. In *E.coli* the *lac* operon encodes sugar metabolizing enzymes depending on the presence of lactose in the medium. In the absence of lactose, repressor proteins bind to the transcription start site of the *lac* operon and inhibit the transcription. Lactose and molecules similar in structure like isopropyl-β-D-thiogalactoside (IPTG) bind to *lac* repressor proteins, which leads to their dissociation from the DNA. Thereby, lactose and IPTG can induce transcription of genes under control of the *lac* operon.

Bacteria were transformed with expression vectors coding for GST-fusion proteins or GST alone. The next day, 9 separate colonies were picked and transferred to 2 mL LB-Amp medium. The colonies were cultured overnight at 37°C. 100 µL of the overnight culture were used to inoculate 2 mL LB-Amp medium. The bacteria grew for 2 h at 37°C. Afterwards 500 µL of each culture were removed as control sample. 1 mM IPTG was added to the suspension to induce the synthesis of the GST-fusion proteins. The bacteria were incubated for 3 h at 37°C. 200 µL of each culture were removed as induced sample. The aliquots were centrifuged for 2 min at 4,000xg (Eppendorf 5417R) at 4°C, and the supernatant was discarded. The samples were denatured in 25 µL of 2x Laemmli loading buffer and analyzed by SDS-Page (6.a). The gel was stained with Coomassie (6.b) to identify inducible colonies.

7.b Large scale purification

7.b.I. Buffers and solutions

Buffer A		1 L
HEPES, pH 7.5	20 mM	20 mL of 1 M stock
NaCl	1 M	58.44 g
DTT	1 mM	10 µL of 1 M stock / 10 mL
PMSF	1 mM	50 µL of 200 mM stock / 10 mL

Na^+- buffer		1 L
NaCl	100 mM	20 mL of 5 M stock
EDTA	1 mM	2 mL of 0.5 M stock
Tris/HCl pH 7.5	20 mM	20 mL of 1 M stock
Nonidet-P40	0.5% (v/v)	5 mL
DTT	1 mM	10 µL 1 M DTT / 10 mL
PMSF	1 mM	50 µL 200 mM PMSF / 10 mL

All buffers were stored at room temperature. DTT and PMSF were added freshly to the buffers before use.

7.b.II. Preparation of glutathione agarose

The amounts of glutathione agarose beads sufficient for a preparation from 1 L bacteria culture depend on affinity properties of the beads stock and were stated in the delivery protocol from Sigma. The respective amount of glutathione agarose beads was pre-swollen overnight in 8 mL PBS at 4°C. The next day, the solution was centrifuged for 3 min at 150xg (Megafuge 1.0) at 4°C. The supernatant was discarded and the beads were washed twice with 10 mL buffer A, followed by 3 min centrifugation at 150xg (Megafuge 1.0) at 4°C. Afterwards, a 50% slurry solution of the beads in buffer A was prepared and kept on ice at 4°C.

7.b.III. Purification by affinity chromatography

After successful screening, 100 mL overnight culture of an inducible clone were used to inoculate 1 L LB-Amp medium. During the growing phase the optical density (OD) at 600 nm wavelength was determined using the UV-visible recording spectrometer UV-160. The bacteria were cultured until an OD_{600} of 0.4-0.6 was reached. At this time point the bacteria were in a logarithmic growing phase. Now, induction of protein synthesis was induced by addition of 1 mM IPTG. After 4 h of incubation, the bacteria were harvested by centrifugation for 15 min at 2,000xg at 4°C, using the Beckmann J2HS centrifuge (rotor JA14). The supernatant was discarded and the pellet was resuspended in 25 mL buffer A. The suspension was frozen immediately using liquid nitrogen and was stored overnight at -80°C. The next day the bacteria were thawed on ice. To destroy the cell walls of the bacteria the suspension was sonicated. The suspension was kept in an ice-methanol bath during the process to prevent overheating and degradation of the proteins. 7 cycles of 20 sec sonication (Branson Sonifyer® Cell Disrupter B15: output control 7, duty cycle 50%) were performed with a pause of 2 min of cooling in ice-methanol between the cycles. Afterwards the cell debris was pelleted by centrifugation for 10 min at 15,500xg at 4°C, using the Beckmann J2HS centrifuge (rotor JA20). The supernatant containing the proteins was transferred to a new 50 mL tube and 1 mL of the 50%-slurry solution of glutathione-agarose beads was added. The supernatant was incubated with the agarose for 3 h, agitating at 4°C (rolling platform TRM-V). Subsequently the agarose was pelleted by centrifugation at 150xg (Megafuge 1.0) at 4°C, and washed 3 times with Na^+-buffer.

Using Na⁺-buffer a 50%-slurry solution of the proteins bound to the agarose was prepared and stored on ice at 4°C.

During the purification process aliquots were taken from the supernatants and finally from the agarose beads. These samples were analysed by SDS-PAGE (6.a) and Coomassie staining (6.b) to control for successful expression and purification of the GST-fusion proteins.

7.c Elution of GST-fusion proteins from the agarose

7.c.l. Buffers and solutions

NaPi buffer		ca. 85 mL
Na_2HPO_4 (basic)	75 mM	70 mL
NaH_2PO_4 (acidic)	75 mM	ca. 15 mL

Dialysis buffer (EMSA)		1 L
HEPES pH 7.5	20 mM	40 mL of 0.5 M HEPES
NaCl	200 mM	40 mL of 5 M NaCl
EDTA	1 mM	2 mL of 0.5 M EDTA
DTT	1 mM	1 mL 1 M DTT
PMSF	1 mM	5 mL 200 mM PMSF

Dialysis buffer (crosslinking)		1 L
NaPi buffer pH 7.0		1 L
DTT	1 mM	1 mL 1 M DTT
PMSF	1 mM	5 mL 200 mM PMSF

Elution buffer pH 7.0 – 7.5		25 mL
Glutathione	50 mM	0.384 g
Dialysis buffer		ad 25 mL

All buffers were stored at room temperature. The pH of the elution buffer was adjusted prior to use to 7.0 – 7.5 using 1 M NaOH. The NaPi buffer was prepared by adjusting the pH of the basic Na_2HPO_4 to 7.0 using the acidic NaH_2PO_4. DTT and PMSF were added freshly to dialysis buffers before use.

7.c.II. Procedure

For electrophoretic mobility shift assays (11.d) and crosslinking experiments (10.b.I) the GST-fusion proteins were eluted from the agarose. To elute GST-fusion proteins needed in crosslinking experiments (10.b.I), the dialysis buffer was modified such that the proteins were already solved in sodium phosphate (NaPi) buffer. The agarose-bound proteins were transferred to a new 2 mL tube and were centrifuged for 3 min at 100xg (Eppendorf 5417R) at 4°C. The supernatant was discarded and 1 vol elution buffer was added to the agarose. The samples were incubated for 5 min at 4°C, agitating at 1,000 rpm (Eppendorf thermomixer). After centrifugation for 3 min at 100xg (Eppendorf 5417R) at 4°C, the supernatant was transferred to a new 1.5 mL tube. This procedure was repeated up to 10 times. To test the concentration of proteins eluted from the agarose a mini-Bradford assay was performed. 5 µL of the supernatant were transferred to 96-well plate. Dye reagent for protein assays (Biorad) was diluted 1:5 in H_2O. 100 µL of the solution was added per well. The Coomassie dye present in the solution indicates high concentration of proteins by a colour change to blue. The extracts containing high protein amount were pooled and transferred to dialysis tubes (GIBCO) and dialysed against 3 L dialysis buffer for 2 times 1.5 h at 4°C. Afterwards the protein extracts were analyzed by Bradford assay and semi-quantitative SDS-PAGE (6.g). Extracts were stored on ice at 4°C. For EMSAs, the purified proteins were diluted as desired using dialysis buffer. To store the extracts at -80°C, 1 vol of 87% glycerol was added.

8. $[^{35}S]$-Labeling of proteins

To analyze the effects of lithium and magnesium on the interaction between CREB and TORC in GST pull-down assays and to investigate effects of lithium on the tetramerization of TORC1, TORC proteins were radioactively labeled with $[^{35}S]$ by *in vitro* transcription and translation. For the *in vitro* labeling, the TNT T7 coupled reticulocyte lysate system according to the manufacturer's instruction was used including T7 polymerase and L-$[^{35}S]$methionine (1000 Ci/mmol) (Hartmann Analytics). The coding sequences for the respective proteins were subcloned into pcDNA3 under control of the T7 promoter.

The following kit components were mixed in a 1.5 mL tube:

Reticulocyte lysate	25 µL
TNT-Buffer	2 µL
Amino acid mix (without Met)	1 µL
Template	1 ng
T7-RNA polymerase	1 µL
L-[^{35}S]Methionine (30 pmol)	3 µL
H$_2$O	ad 50 µL

The mixture was incubated for 90 min at 30 °C agiating at 400 rpm (Eppendorf thermomixer).

To control for successful *in vitro* transcription / translation 1 µL and 0.5 µL aliquots were electrophoresed by SDS-PAGE (6.a). The gel was dried and the radioactively labeled proteins were analysed by phosphor-imaging (6.f).

9. GST pull-down assay

9.a Buffers and solutions

Reaction buffer		100 mL
NaCl	100 mM	2 mL of 5 M NaCl
EDTA	1 mM	0.2 mL of 0.5 M EDTA
Tris/HCl pH 7.5	20 mM	2 mL 1 M Tris/HCl pH 7.5
Nonidet-P40	0.5 % (v/v)	0.5 mL
DTT	1 mM	10 µL 1 M DTT / 10 mL
PMSF	1 mM	50 µL 200 mM PMSF / 10 mL

Depending on the experiment, buffers contained varying concentrations of LiCl or MgCl$_2$. To avoid changes in ionic strength the total concentration of salt was kept at 100 mM and was balanced against NaCl.

Salt composition in buffers used in GST pull-down assays:

NaCl	MgCl$_2$	LiCl
100 mM	-	-
99.5 mM	-	0.5 mM
95 mM	-	5 mM
80 mM	-	20 mM
99 mM	0.5 mM	-
90 mM	5 mM	-
80 mM	10 mM	-
60 mM	20 mM	-
95 mM	-	5 mM
94 mM	0.5 mM	5 mM
85 mM	5 mM	5 mM
75 mM	10 mM	5 mM
55 mM	20 mM	5 mM

All buffers were stored at room temperature. DTT and PMSF were added freshly to the reaction buffers before use.

9.b Assay conditions

In general, GST pull-down assays are used to test *in vitro* the interaction between proteins in a cell-free condition. One putative interaction partner is expressed as recombinant GST-fusion protein and is immobilized on glutathione agarose. The other protein is labeled radioactively. The amount of radioactively-labeled protein bound to the recombinant protein is detected by analysis of the protein mixture using autoradiography. Comparison to a reference (negative control, e.g. GST alone) is needed to determine the specificity of the interaction. In the present work, GST pull-down assays were performed to investigate effects of lithium and magnesium on the interaction between GST-fusion proteins of CREB and [^{35}S]-labeled TORC proteins. The method was perfomed as previously described (Kratzner et al., 2008). GST-fusion proteins of CREB were expressed in *E.coli* and purified by affinity chromatography (7.b). [^{35}S]-Labeling of TORC proteins was performed by *in vitro* transcription / translation using [^{35}S]methionine (8.). Equal amounts of GST-fusion proteins and GST alone were determined by SDS-PAGE (6.a) followed by Coomassie staining (6.b). The respective amounts of agarose were transferred to a 0.5 mL tube and were washed in 200 µL of reaction buffer. After centrifugation for 2 min at 100xg (Eppendorf 5417R) at 4°C, the supernatants were discarded. The agarose-bound proteins were supplemented with sufficient amount of reaction buffer and 3.5 µL of [^{35}S]-labeled protein to perform the reaction in a final volume of 250 µL. The binding reaction was carried out at 4°C agitating on a wobbling disk (Polymax 1040) for 20 h. The next day, the samples were centrifuged at 60xg (Biofuge 15R) at 4°C, and the supernatant was discarded. The agarose was washed 3 times with 400 µL of reaction buffer followed by centrifugation for 2 min at 60xg (Biofuge 15R) at 4°C, and discard of the supernatant. Afterwards the agarose was denatured in 20 µL of 2x Laemmli-loading buffer and electrophoresed by SDS-PAGE (6.a). The gels were dried and analyzed by autoradiography (6.f).

10. Crosslinking of proteins using glutaraldehyde

In the present study the effect of lithium on the oligomerization of isolated TORC1 was tested by means of glutaraldehyde cross-linking. For this approach, either the first 44 amino acids of TORC1 fused to GST (GST-TORC1$_{1-44}$), or the first 327 amino acids of TORC1 labeled radioactively with [^{35}S] ([^{35}S]TORC1$_{327}$) were employed.

10.a Buffers and solutions

NaPi buffer		ca. 85 mL
Na$_2$HPO$_4$ (basic)	75 mM	70 mL
NaH$_2$PO$_4$ (acidic)	75 mM	ca. 15 mL

0.25% Glutaraldehyde		10 mL
Glutaraldehyde	0.25% (v/v)	100 µL of 25% Glutaraldehyde
H$_2$O		ad 10 mL

LiCl		10 mL
LiCl	200 mM	1 mL of 2 M LiCl
H$_2$O		ad 10 mL

NaCl		*10 mL*
NaCl	200 mM	400 µL of 5 M NaCl
H$_2$O		ad 10 mL

All buffers and solutions were stored at room temperature. The NaPi buffer was prepared by adjusting the pH of the basic Na$_2$HPO$_4$ to 7.0 using the acidic NaH$_2$PO$_4$.

10.b Reaction conditions

To link proteins to each other, 0.01% (v/v) glutaraldehyde were used. The crosslinking effect is dominated by reactions with the ε-amino groups of lysine residues of proteins in close proximity (Wine et al., 2007). The reaction was performed as described before with some modifications (Conkright et al., 2003a).

10.b.I. Crosslinking of GST-TORC1$_{1-44}$

GST-fusion proteins were expressed in *E.coli*, purified by affinity chromatography, and were eluted from the glutathione agarose (7.a – c). To control for unspecific crosslinking due to the GST-tag, GST alone was used.

The following components were mixed in a 1.5 mL tube:

Protein	4 µg
LiCl / NaCl	20 mM
Glutaraldehyde	0.01% (v/v)
NaPi buffer	ad 20 µL

The samples were incubated for 1 h at 20°C, agitating at 400 rpm (Eppendorf thermomixer). Afterwards sufficient amounts of 2x Laemmli-loading buffer were added, the samples were denatured and the proteins were separated by SDS-PAGE. Protein staining on the gel was performed by silver staining (6.c).

10.b.II. Crosslinking of [^{35}S]TORC1$_{327}$

TORC1$_{327}$ was radioactively labeled by *in vitro* transcription / translation using [^{35}S]methionine (8.). The following components were mixed for the reaction in a 1.5 mL tube:

[^{35}S]-labeled protein	5 µL
LiCl / NaCl	20 mM
Glutaraldehyde	0.01%
NaPi buffer	ad 20 µL

The crosslinking was performed for 30 min at 20° C, agitating at 400 rpm (Eppendorf thermomixer). The samples were denatured with sufficient amount of 2x Lämmli-loading buffer and analysed by SDS-PAGE followed by radiography (6.f)

11. Electrophoretic mobility shift assay

Electrophoretic mobility shift assays (EMSAs) are used to investigate DNA-protein interactions. Non-denaturing conditions are chosen for electrophoresis to ensure naïve conformation of the proteins. Protein-bound DNA is migrating slower than free DNA visible by a band shift after radiography. In the present work, EMSAs were used to investigate the DNA-binding ability of the CREB mutant CREB-K290E. Furthermore the occurrence of both CREB and TORC1 in complex with the CRE of the rat *somatostatin* gene promoter was investigated by supershift experiments (11.d.II).

11.a Buffers and solutions

TBE	5x	2 L
Tris	450 mM	108.9 g
Boric acid	450 mM	55.6 g
EDTA pH 8.0	10 mM	40 mL 0.5 M EDTA

Acrylamide for Shift-Gel	40% (w/v)	100 mL
Acrylamide	38% (w/v)	38 g
Bis-acrylamide	2% (w/v)	2 g

Stop mix blue for EMSA		10 mL
Glycerol	30%(v/v)	3.39 mL 87% Glycerol
Bromophenolblue	0.25% (w/v)	0.025 g
Xylene cyanol FF	0.25% (w/v)	0.025 g
H_2O		ad 10 mL

Binding buffer		500 µL
Tris / HCl pH 7.5	50 mM	25 µL 1 M Tris/HCl pH 7.5
NaCl	100 mM	50 µL 1 M NaCl
Glycerin	10% (v/v)	100 µL 50% Glycerin
$MgCl_2$	20 mM	10 µL of 1 M $MgCl_2$-Sol.
BSA	12 µg / mL	6 µL of 1 mg / mL BSA-Sol.
poly(dI-dC)	80 mg / mL	40 µL of 1 µg /µL stocksol.

All buffers were stored at room temperature. The acrylamide solution was stored at 4°C. BSA and poly(dI-dC) were added freshly before use of the binding buffer.

11.b Gel preparation

For EMSAs non-denaturing gels were used which are free from SDS.

For a 5% polyacrylamide gel the following components were mixed:

Acrylamide	5% (w/w)	6.25 mL of acrylamide for shift gels (40%)
TBE	0.5 x	5 mL 5 x TBE
H_2O		38 mL
TEMED		50 µL
APS		500 µL 10% APS

The gel solution was prepared, cast between 2 glass electrophoresis plates (18 cm x 16 cm x 3 mm), and the comb forming the slots was inserted immediately. After polymerization the gel was placed in the electrophoresis chamber SE 600. The chamber was filled with 0.5 x TBE buffer and a pre-electrophoresis was performed for 1 h at 180 V to prewarm the gel.

11.c Labeling of oligonucleotides

11.c.I. Annealing

To generate double-stranded probes, complementary oligonucleotides were used. The oligonucleotides were designed such, that the probe contains a 5'-GATC overhang after annealing (Oetjen et al., 1994). For the reaction the following components were mixed in a 0.5 mL tube:

Oligonucleotide forward	100 pmol
Oligonucleotide reverse	100 pmol
NaCl	10 mM
H$_2$O	ad 50 µL

The tubes were placed in water of 85°C temperature which was cooled down to room temperature overnight.

11.c.II. Labeling with [^{32}P]

The probes were labeled by a fill-in reaction using [α-^{32}P]-dCTP (3000 Ci/mmol, Hartmann Analytics) and Klenow enzyme. For the reaction the following components were used:

Double-stranded probe	10 pmol
dATP	1 mM
dGTP	1 mM
dTTP	1 mM
[α-^{32}P]-dCTP	20 µCi
Klenow buffer	1 x
Klenow fragment	2 U
H$_2$O	ad 20 µL

The reaction was performed for 30 min at 25°C (Eppendorf thermomixer).

11.c.III. Purification of the labeled probe

The labeled probe was purified from the free [α-^{32}P]-dCTP by gel filtration chromatography using Mini Quick Spin™ Oligo Columns (Roche) according to the manufacturer's instruction. The Sephadex matrix was homogenized by vortexing and flipping. The storage buffer inside the column was removed by centrifugation for 1 min at 1,000x*g* (Biofuge 15R) at 4°C. The column was placed in a new 1.5 mL tube and the probe was carefully pipetted on the matrix. The purification was performed by centrifugation for 1 min at 1,000xg (Biofuge 15R) at 4°C. The eluate contained the purified probe which was quantified immediately.

11.c.IV. Quantification of the incorporation

1 µL of the probe was mixed with 4 mL of scintillation suspension and placed in a beta counter (LS1801, Beckmann). Depending on the counts per min (cpm) detected, reflecting the quality of incorporation of [α-^{32}P]-dCTP, the probe was diluted to a concentration of 20,000 cpm per µL. The probe was stored at -20°C.

11.d Binding reaction

11.d.I. General procedure

In the present study, GST-fusion proteins of CREB wild-type and mutant K290E were analyzed with respect to their ability to bind to the CRE of the rat *somatostatin* gene. The reaction was performed with some modification as described before (Knepel et al., 1990). Defined amounts of proteins were incubated in binding buffer for 10 min on ice. The binding buffer contains poly(deoxyinosinic-deoxycytidylic) acid [poly(dI-dC)], which is a polymer acting as unspecific competitor DNA. Afterwards the amount of labeled probe corresponding to 20,000 cpm was added and incubated on ice for 15 min. Subsequently, the desired amount of stop mix was added and the samples were loaded on the gel.

11.d.II. Supershift

To verify the presence of a certain protein in a protein-DNA complex, the proteins are incubated with a specific antibody against the protein. Since the electrophoresis is performed under non-denaturing conditions, the protein-DNA complex is expected to migrate differently when an antibody is bound to the protein. In the present work, the CREB-KID antibody was used to detect GST-CREB proteins in complex with DNA, whereas the panTORC antibody was employed to detect TORC. The respective amounts of antibody are given in table 9 (3.f). To perform the supershift, the proteins were incubated for 20 min at room temperature in binding buffer without poly(dI-dC) but containing the respective amounts of antibody. Afterwards the mixture was stored on ice at 4°C. The next day, poly(dI-dC) and the labeled probe were added and left for 15 min on ice. The samples were loaded immediately with sufficient amount of stop mix and electrophoresed.

11.e Electrophoresis

The slots of the gel were washed carefully with TBE buffer. The samples were loaded on the gel and a constant power of 180 V was applied to the gel (Biometra® Standard Power Pack P25). Electrophoresis was performed until the lowest band from the blue Stop mix had passed ¾ of the gel. Afterwards the gel was dried and analyzed as described (section 6.f).

Materials and methods

12. Transient transfection of HIT-T15 cells

In the present work HIT-T15 cells were transiently transfected by use of the DEAE-Dextran method for luciferase reporter-gene assays. Lipofection using Metafectene was applied in all other experiments requiring the transient transfection of expression plasmids. DEAE-Dextran is a cationic polymer which forms a complex with DNA and facilitates endocytotic uptake of the DNA by a mechanism not understood so far. The method of lipofection is 5 – 100-fold more effective compared to transfection by DEAE-Dextran. The principle is the formation of lipid-DNA complexes with complete entrapment of the DNA and fusion with the cell membrane by a mechanism not completely understood (Felgner et al., 1987).

12.a Buffers and solution

TD-buffer		500 mL
Tris/HCl pH 7.4	25 mM	12.5 mL of 1 M Tris/HCl pH 7.4
NaCl	140 mM	14 mL of 5 M NaCl
KCl	5 mM	1.25 mL of 2 M KCl
K_2HPO_4	0.7 mM	0.7 mL of 0.5 M K_2HPO_4

DEAE-Dextran solution		10 mL
DEAE-Dextran	60 mg / mL	600 mg
H_2O		ad 10 mL

The TD buffer was autoclaved for 20 min at 120°C and was afterwards stored at room temperature. The DEAE-Dextran solution was filtrated by a 0.45 µm syringe filter and stored at 4°C.

12.b Transfection using DEAE Dextran

The transient transfection was performed as previously described with some modifications (Sambrook et al., 1989). One 15-cm culture dish with cells at full confluence (~35 x 10^6 cells) was used which is sufficient to seed cells for twelve 6 cm dishes at a density of ~3 x 10^6 cells per dish. The medium was removed and the cells were washed carefully with PBS. Subsequently the cells were removed from the plate by treatment with 3 mL trypsin/EDTA solution (GIBCO) at 37°C for 3 - 5 min. The cells were collected in 7 mL RPMI complete medium, transferred to a 50 mL tube and centrifuged for 2 min at 300xg (Megafuge 1.0) at room temperature. The supernatant was discarded and the cells were washed once with 10 mL TD buffer to remove remnants of trypsin and medium. After centrifugation for 2 min at 300xg (Megafuge 1.0), the cells were resuspended carefully in 1 mL TD buffer per 6 cm dish. 5 µL DEAE-Dextran solution per dish were added followed by addition of 2 µg expression plasmid per dish. When two or more expression plasmids were cotransfected, the total amount of DNA was kept constant by addition of the empty vector pBluescript. The cells were incubated in suspension for 20 min at room temperature. After centrifugation for 2 min at 300xg (Megafuge 1.0) the supernatant was discarded and the cells were washed once with RPMI complete medium. The cells were centrifuged again for 2 min at 300xg (Megafuge 1.0) and the supernatant was removed. The cells were resuspended in 5 mL RPMI complete medium per dish and were seeded immediately.

12.c Transfection using Metafectene

For transfection by lipofection, Metafectene (Biontex) was used. Cells were cultured as appropriate for the experiment in 6-cm dishes or 15-cm dishes and were seeded at a density of ~0.08 x 10^6 cells per cm² or 0.1 x 10^6 cells per cm², respectively. The transfection was performed according to the recommendations of the manufacturer. Metafectene and the DNA were each mixed with pure RPMI. The ratio of volume Metafectene in µL to the amount of DNA in µg was at 2:1.

Amounts of reagents used for the transfection:

Per 15-cm dish:

DNA in 500 µL RPMI + Metafectene in 500 µL RPMI → mix (total volume / dish: 1 mL)

Per 6-cm dish:

DNA in 200 µL RMPI + Metafectene in 200 µL RPMI → mix (total volume / dish: 400 µL)

The mixture was incubated for 15 min at room temperature and subsequently added to the cells during the first hour after seeding.

13. Treatment of HIT-T15 cells

In the present study HIT-T15 were used in different luciferase reporter gene assays, immunocytochemistry experiments, and ChIP assays. To analyze effects upon elevated calcium levels, the cells were treated with KCl leading to membrane depolarization of HIT-T15 cells and opening of voltage-gated calcium channels of the L-type accompanied by an increase in the intracellular calcium concentration. The treatment of cells with forskolin (FSK) inceased the intracellular cAMP level as FSK is a potent activator of the adenylyl cyclase. 8-bromo-cAMP is a membrane permeable cAMP analog which allows an increase in intracellular cAMP levels without activation of the AC. Additionally two different drugs were used in the present study. Cyclosporin A (CsA) is an immunosuppressive agent potently inhibiting the calcium/calmodulin-dependent phosphatase calcineurin. Lithium chloride (LiCl) was used to investigate effects of lithium on cAMP-induced CREB-dependent gene transcription.

In luciferase reporter-gene assays, cells were treated with 1 mM or 2 mM 8-bromo-cAMP for 6 h, with 20 mM LiCl for 7 h, or with the combination of 8-bromo-cAMP and LiCl. The teatment with LiCl was initated 1 h in advance to the treatment with 8-bromo-cAMP. Additionally cells were treated with 45 mM KCl for 6 h, with 10 µM FSK for 6 h, or with the combination of FSK and KCl.

For the interaction analysis by the mammalian two hybrid assay HIT-T15 cells were treated for 6 h with 1 mM 8-bromo-cAMP, for 30 h with 20 mM LiCl or with the combination of LiCl and 8-bromo-cAMP.

For immunocytochemistry experiments, HIT cells were treated for 30 min with 1 mM or 2 mM 8-bromo-cAMP, for 90 min with 20 mM LiCl, or with the combination of 8-bromo-cAMP and LiCl. The treatment with LiCl was started 1 h before treatment with 8-bromo-cAMP. Additionally, cells were teated with 45 mM KCl for 30 min, with 5 µM CsA, or with the combination of KCl and CsA. The treatment with CsA was started 1 h prior to treatment with KCl.

For ChIP assays, HIT-T15 cells were treated for 30 min with 2 mM 8-bromo-cAMP, for 90 min with 20 mM LiCl, or with the combination of 8-bromo-cAMP and LiCl. Treatment with LiCl was started 1 h in advance to treatment with 8-bromo-cAMP.

14. Preparation of whole-cell extracts (hot lysis)

The general expression level of GAL4-fusion proteins of CREB was examined by Western blot in the present work. For the analysis, whole-cell protein extracts were used and prepared by hot lysis. For that purpose, the cells were cultured in 6-cm dishes and transfected with the respective expression plasmid using Metafectene. 48 h after seeding the medium was discarded and the cells were rinsed with 3 mL PBS. The cells were detached from the dish with a scraper and were collected in 150 µL 90°C hot 2x Laemmli-loading buffer. The suspension was sheared 5 times using a 1-mL syringe (needle 24G). Subsequently the samples were incubated at 99°C for 10 min and the expression level of GAL4-CREB proteins was analyzed by Western blot using the CREB-KID anitbody (6.d; table 9).

15. Preparation of cytosolic and nuclear extracts

15.a Buffers and solutions

Digitonin solution		50 mL
Digitonin	0.01% (w/v)	5 mg
EDTA	1 mM	100 µL 0.5 M stock
Protease inhibitors	1x	200 µL of 50 x Protease inhibitor mix

Triton solution		50 mL
Triton X-100	2% (w/v)	1 mL
EDTA	1 mM	100 µL 0.5 M stock
Protease inhibitors	1x	200 µL of 50 x Protease inhibitor mix

Buffer C (Schreiber et al., 1989)		50 mL
HEPES pH 7.9	20 mM	1 mL 1M stock
NaCl	400 mM	4 mL 5M stock
EDTA	1 mM	100 µL 0.5 M stock
EGTA	1 mM	278 µL 180 mM stock
DTT	1 mM	50 µL 1 M DTT
PMSF	1 mM	250 µL 200 mM PMSF

Protease	Inhibitor	
Mix	50 x	515 µL
PMSF	50 mM	500 µL
Aprotinin	250 µg	5 µL of 50 µg / µL stock
Pepstatin	250 µg	5 µL of 50 µg / µL stock
Leupeptin	250 µg	5 µL of 50 µg / µL stock

All buffers and solutions were stored at 4°C. The 50x protease inhibitor mix was stored at -20°C. Digitonin solution and Triton solution were supplemented with protease inhibitors freshly before use. DTT and PMSF were added to the buffer C freshly before use.

15.b Extraction

In the present work, the expression level of VP16-fusion proteins of the bZip wt and different mutants was exmined by Western blot. At the same time, the localization of the proteins was of interest. For that purpose, cytosolic and nuclear extracts were prepared by a combination of the differential detergent fractionation (Ramsby et al., 1994) and the extraction of nuclear proteins (Schreiber et al., 1989). HIT-T15 cells were cultured on 15-cm dishes and transiently transfected using Metafectene (12.c). 48 h after transfection the medium was removed and the cells were rinsed carefully with 10 mL PBS. To extract the cellular protein fraction, digitonin was used in a first step. Digitonin is a mild detergent perforating the cell membrane. Here, 750 µL of digitonin solution were dropped onto the cells which were mechanically detached from the dish with a scraper. The suspension was transferred to a 1.5 mL tube and lysed on ice for 5 min. The samples were centrifuged for 1 min at 16,000xg (Eppendorf 5417R) at 4°C. The supernatant contained the cytosolic fraction. The resulting pellet was resuspended in Triton solution. Triton X100 is a detergent solubilizing the lipid membrane. In contrast to SDS it does not denature the proteins, therefore it can be used to extract membrane-bound proteins. Incubation in Triton solution was performed for 30 min on ice. After centrifugation for 5 min at 16,000xg (Eppendorf 5417R) at 4°C, the supernatant contained membrane proteins and the organelle fraction. This fraction was not used for the analysis and was therefore discarded. The pellet was now supplemented with 50 µL of buffer C. Solubilization of the nuclear proteins was performed 15 min at 4°C, agitating at 1,400 rpm (Eppendorf thermomixer). The samples were centrifuged for 5 min at 20,000xg (Eppendorf 5417R) at 4°C. The supernatant consisted of the nuclear protein fraction. The protein concentration of the extracts was determined photometrically (6.g.III) and the expression levels of VP16-fusion proteins were analyzed by Western blot using the VP16 (1-21) antibody (6.d; table 9).

16. Immunocytochemistry

16.a Buffers and solutions

Quenching buffer		50 mL
Sodium borohydride	0.1% (w/v)	0.05 g
PBS		50 mL

Blocking buffer		50 mL
BSA	1% (w/v)	0.5 g
Horse Serum	10% (v/v)	5 mL
PBS		44.5 mL

Antibody-dilution buffer		1.4 mL
BSA	1% (w/v)	0.014 g
PBS		1.4 mL

All buffers were prepared freshly before use.

16.b Staining procedure

In the present study the translocation of endogenous TORC proteins in response to treatment with KCl, CsA, 8-bromo-cAMP, or LiCl was analyzed by immunocytochemistry. For that purpose HIT-T15 cells were grown on coverslips in 6-well plates (0.962cm² per well) and seeded at a density of ~1.8×10^5 cells per cm². The cells were treated as desired 48 h after seeding (13.). The procedure was performed as described before (Plaumann et al., 2008). The coverslips were transferred to a new 6-well plate and the cells were washed once with 4 mL PBS. The PBS was aspirated off completely and the cells were fixated by immersion for 10 min in 100% methanol (-20°C) to precipitate the proteins. The coverslips were rinsed with PBS 3 times for 5 min, agitating at room temperature (Titramax 100). The self-fluorescence of the cells was reduced by incubation in quenching buffer for

5 min, agitating at room temperature (Titramax 100). The coverslips were washed in PBS 3 times for 5 min, agitating at room temperature (Titramax 100). To block unspecific binding of the antibody, the coverslips were treated with blocking buffer for 45 min at room temperature. For the staining procedure, the primary antibody panTORC was diluted 1:1500 in antibody-dilution buffer (table 9). Incubation was performed on parafilm in a cell culture dish. Here, 50 µL of the antibody dilution was put on the parafilm for each coverslip. The coverslips were put on the antibody dilution such that the cells on the coverslip were in direct contact to the solution. Incubation was overnight at 4°C. To prevent drying of the cells, some paper soaked with PBS were kept in the dish together with the coverslips. The following day, the coverslips were transferred to a 6-well plate and rinsed with PBS 3 times for 5 min, agitating at room temperature (Titramax 100). The secondary rabbit IgG antibody linked to the fluorophore Alexa488® was diluted in antibody-dilution buffer (table 9). The incubation was again performed on parafilm using 50 µL of antibody dilution per coverslip. The staining was carried out for 30 min at room temperature. To prevent bleaching of the secondary antibody, the coverslips were kept in the dark. Afterwards the cells on the coverslips were washed with PBS 3 times for 5 min at room temperature and low lighting conditions. For microscopy, the coverslips were mounted on slides using the Vectashield® Mounting Medium with DAPI (4',6-diamidino-2-phenylindol). DAPI is a dye that intercalates with the DNA and thereby allows the discrimination of the nucleus from the cytoplasm during fluorescence microscopy.

16.c Fluorescence microscopy

The cells were observed at 63x magnification (Zeiss, LD Achroplan 63x) using the Zeiss Axiovert 200 microscope. The Software OpenLab™ 3.1 (Improvision Ltd.) was used to operate the device. Excitation wavelengths for the fluorophores were: 495 nm for AlexaFluor® 488 which emits at 519 nm and 360 nm for DAPI which emits light at 460 nm wavelength.

17. Chromatin-immunoprecipitation (ChIP)

17.a Buffers and solutions

Cell lysis buffer		100 mL
Tris/HCl pH 8.0	10 mM	1 mL of 1 M stock
NaCl	10 mM	200 µL of 5 M stock
Nonidet-P40	0.2% (v/v)	2 mL of 10% Nonidet-P40
Protease inhibitors	1x	200 µL of 50x protease inh. mix / 10 mL

Nuclei lysis buffer		100 mL
Tris/HCl pH 8.0	50 mM	5 mL of 1 M stock
EDTA	10 mM	2 mL of 0.5 M stock
SDS	1% (w/v)	10 mL of 10% SDS
Protease inhibitors	1x	200 µL of 50x protease inh. mix / 10 mL

Wash buffer I		50 mL
Tris/HCl pH 8.0	20 mM	1 mL of 1 M stock
NaCl	150 mM	1.5 mL of 5 M stock
EDTA	2 mM	0.2 mL of 0.5 M stock
SDS	0.1%	0.5 mL of 10% SDS
Triton X100	1% (v/v)	0.5 mL

Wash buffer II		50 mL
Tris/HCl pH 8.0	20 mM	1 mL of 1 M stock
NaCl	500 mM	5 mL of 5 M stock
EDTA	2 mM	0.2 mL of 0.5 M stock
SDS	0.1%	0.5 mL of 10% SDS
Triton X100	1% (v/v)	0.5 mL

Wash buffer III		50 mL
Tris/HCl pH 8.0	10 mM	0.5 mL of 1 M stock
EDTA	1 mM	0.1 mL of 0.5 M stock
LiCl	0.25 M	3.125 mL of 4 M stock
Nonidet-P40	1% (v/v)	0.5 mL
Deoxycholic acid	1% (w/v)	500 mg

Elution buffer I		10 mL
Tris/HCl pH 8.0	10 mM	100 µL of 1 M stock
EDTA	1 mM	20 µL of 0.5 M stock
SDS	1% (w/v)	1 mL of 10% SDS

Elution buffer II		10 mL
Tris/HCl pH 8.0	10 mM	100 µL of 1 M stock
EDTA	1 mM	20 µL of 0.5 M stock
SDS	0.67% (w/v)	670 µL of 10% SDS

TE buffer		50 mL
Tris/HCl pH 8.0	10 mM	0.5 mL of 1 M stock
EDTA	1 mM	0.1 mL of 0.5 M stock

Protease inhibitor mix	50 x	515 µL
PMSF	50 mM	500 µL
Aprotinin	250 µg	5 µL of 50 µg / µL stock
Pepstatin	250 µg	5 µL of 50 µg / µL stock
Leupeptin	250 µg	5 µL of 50 µg / µL stock

All buffers and solutions were stored at room temperature. The 50x protease inhibitor mix was stored at -20° C. Protease inhibitors were added freshly from the 50x solution to cell lysis solution and nuclei lysis solution before use.

17.b ChIP

Chromatin immunoprecipitation (ChIP) is a method widely used to study protein:DNA interactions. It involves the crosslinking of proteins and DNA by formaldehyde followed by immunoprecipitation using specific antibodies. DNA analysis after precipitation allows characterization of native binding sites of transcription factors (Wells and Farnham, 2002). In the present work, this method was used to investigate effects of lithium and 8-bromo-cAMP on the recruitment of the transcription factor CREB and its cofactor TORC1 to different promoters. After crosslinking with formaldehyde the complex was precipitation by use of TORC1 antibody. The amount of precipitated promoter DNA was quantified afterwards to consider changes in the interaction between the transcription factor and its cofactor at the promoter.

HIT-T15 cells were grown in 6-cm dishes (28.27cm² surface) and seeded at a density of ~0.08 x 10^6 cells per cm². The cells were transfected using Metafectene (Biontex). 48 h after transfection the cells were treated as desired (13.). Each group was performed in duplicate and the lysates were pooled at the indicated time point. The ChIP assay was performed as described before (Kratzner et al., 2008). The proteins were cross-linked to each other and to DNA by incubation of the cells with 1% (v/v) formaldehyde for 20 min, agitating at room temperature on a rocking platform. The process was stopped by addition of 125 mM glycine and incubation for 5 min, agitating at room temperature on a rocking platform. Afterwards the medium was discarded and the cells were washed twice with ice-cold PBS. The cells were scraped in 1.5 mL PBS and centrifuged for 2 min at 1,700x*g* (Eppendorf 5417R) at 4°C. The pellet was resuspended in cell lysis buffer, freezed in liquid nitrogen and thawed in a waterbath at 37°C. The samples were homogenized using a 1 mL dounce homogenizer (Kontes Glas) and incubated 10 min on ice. After 5 min centrifugation at 2,700x*g* (Eppendorf 5417R) at 4°C, the pellet was resuspended in nuclei lysis buffer, the lysates of the two samples per group were pooled and incubated on ice for 10 min. The samples were sonicated briefly to shear the DNA (Branson Sonifyer® Cell Disrupter B15: output control 5, duty cycle 50%) and centrifuged for 10 min at 18,000x*g* (Eppendorf 5417R) at 4°C. The supernatant was transferred to a new tube. At that stage, 10% of the lysate was taken as input control. To reduce unspecific binding of proteins to the agarose, the lysate was precleared by incubation with 25 µL of Sepharose CL-4B (50%-slurry solution, equilibrated to nuclei lysis buffer plus 0.1% [w/v] BSA) for 3 h, rotating at 4°C (GFL 3025). The FLAG M2 antibody for immunoprecipitation (table 9) was prebound to protein-G agarose (50%-slurry solution, equilibrated to nuclei lysis buffer with 0.1% [w/v] BSA) for 1 h, rotating at 4°C (GFL 3025). 25 µL of antibody bound to protein-G agarose were added per sample and incubated overnight, rotating at 4°C (GFL 3025). The next day the samples were centrifuged 1 min at 18,000xg (Eppendorf 5417R) at 4°C. The supernatant was discarded. The agarose was washed twice with wash buffer I, once with wash buffer II, and once with wash buffer III. After each step the samples were centrifuged 2 min at 1,700xg (Eppendorf 5417R) at 4°C, and the supernatant was discarded. Next, the proteins were eluted from the agarose by incubation of the pellet in elution buffer I for 15 min at 65°C. After centrifugation for 3 min at 18,000xg (Eppendorf 5417R) at 4°C, the supernatant was transferred to a new tube. The pellet was now incubated with elution buffer II for 15 min at 65°C. Again the samples were centrifuged 3 min at 18,000xg (Eppendorf 5417R) at 4°C and the supernatant was transferred to the tube containing already the supernatant from elution step I. For the input control, 20 µL 10% SDS were added to the samples and the volume was filled up to 250 µL with TE buffer. All samples and the input controls were incubated overnight at 65°C. The following day, 250 µL TE buffer and 100 µg Proteinase K were added to each sample and incubated 2 h at 37°C to digest the proteins in the samples. Addition of 400 mM LiCl precipitated high molecular RNA. DNA purification was performed by phenol – chloroform / isoamylalcohol extraction: 500 µL of phenol were added to the sample, mixed and centrifuged for 2 min at 16,000xg (Biofuge pico) at room temperature. The aqueous phase was transferred to a new tube. This procedure was repeated once with phenol and once with 500 µL chloroform / isoamylalcohol (24:1). The DNA was precipitated by addition of 900 µL 100% ethanol (-20°C) and 10 µg glycogen, and centrifuged for 35 min at 20,800xg (Eppendorf 5417R) at 4°C. The supernatant was discarded and the pellet washed with 500 µL 70% ethanol. The samples were centrifuged 10 min at 20,800xg (Eppendorf 5417R) at 4°C. The resulting pellet was dried and solved in 20 µL TE buffer. The samples were analyzed by quantitative real-time PCR.

17.c Quantitative real-time PCR

To analyze samples from ChIP assays quantitatively, a real-time PCR was performed using TaqMan™ probes. The TaqMan™ probe is an oligonucleotide labeled with a fluorescent dye at its 5' end and contains a quencher at its 3' end. Thus fluorescence-energy transfer reduces fluorescence from the probe (Figure 6, top). The probe is disabled from 3' extension by a 3' phosphate and is designed such that it hybridizes to a specific sequence in the DNA template to be quantified. During amplification of the target sequence, defined by specific primers, the probe hybridized to the DNA is digested due to the 5' → 3' exonuclease activity of the polymerase (Figure 6, bottom). As soon as the nucleotides are separated the indicator fluorescence is restored (Lee et al., 1993). Spectrophotometric measurement of the fluorescence during PCR gives a sensitive method to determine amounts of defined DNA present in the sample. This method was used to determine the amounts of DNA precipitated in ChIP assays (16.b).

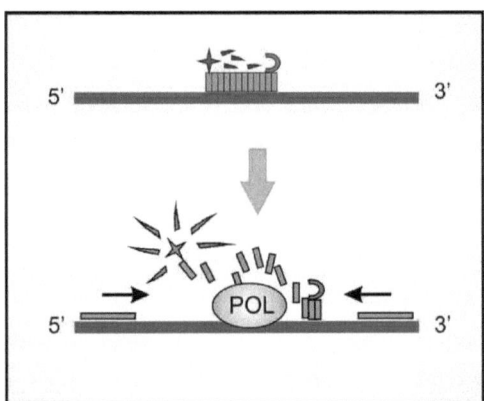

Figure 6: Principle of quantitative real-time PCR using TaqMan™ probes.
The TaqMan™ probe is labeled with a fluorescent dye at its 5' end and with a quencher at the 3' end. Due to fluorescence energy transfer the fluorescence is reduced as long as the probe is intact (top). The probe anneals to the sequence to be amplified. The target sequence is defined by specific primers. During PCR the 5' → 3' exonuclease activity of the polymerase separates the nucleotides of the probe and thereby restores the indicator fluorescence.

A calibration curve was generated as reference using the pure plasmid. The plasmid was diluted to 10^{10} molecules per µL. This solution was used for a dilution series in steps of 10 to get samples with $10^8 - 10^2$ molecules per µL, which are used as reference samples. The real-time PCR was performed using the TaqMan® Gene Expression Master Mix including the AmpliTaq Gold® DNA Polymerase (Applied Biosystems). Each sample was measured in triplicate on a 384-well plate using the ABI PRISM 7900 HT Sequence Detection System (Applied Biosystems). The following reaction mix was prepared per sample:

2x Master Mix	1x	9 µL
Primer forward	16 pmol	0.16 µL of 100 µM primer
Primer reverse	16 pmol	0.16 µL of 100 µM primer
TaqMan™ probe	1.8 pmol	0.36 µL of 5 µM TaqMan™ probe
DNA		3 µL
H_2O		ad 18 µL

The conditions for the PCR regarding duration and temperature of annealing and elongation were tested beforehand for the different pairs of primers and probes, and were adjusted as appropriate. To analyze the raw data the software SDS 2.1 (Applied Biosystems) was used.

18. Luciferase reporter-gene assay

18.a Buffers and solutions

Scraping buffer		500 mL
Tris/HCl pH 7.5	40 mM	20 mL of 1 M Tris/HCl pH 7.5
EDTA	1 mM	1 mL of 0.5 M EDTA pH 8.0
NaCl	150 mM	15 mL of 5 M NaCl

KPi buffer		ca. 540 mL
K_2HPO_4 (basic)	100 mM	500 mL
KH_2PO_4 (acidic)	100 mM	ca. 40 mL

Glycylglycine buffer		500 mL
Glycylglycine pH 7.8	25 mM	24 mL of 0.5 M glycylglycine pH 7.8
MgSO$_4$	15 mM	7.5 mL of 1 M MgSO$_4$
EGTA	4 mM	11.1 mL of 180 mM EGTA

Luciferin stock		180 mL
DTT	10 mM	275 mg
Luciferin	1 mM	50 mg
Glycylglycine buffer		178.5 mL

Luciferase-assay mix		for 25 samples
Glycylglycine buffer		7.5 mL glycylglycine buffer
KPi buffer	16.5 mM	1.5 mL KPi buffer
ATP	2 mM	100 µL of 200 mM ATP
DTT	1 mM	10 µL of 1 M DTT

Luciferin solution		for 25 samples
Glycylglycine buffer		5.6 mL glycylglycine buffer
DTT	10 mM	56 µL of 1 M DTT
Luciferin	250 µM	1.4 mL luciferin stock

KPi buffer was prepared by adjusting the pH of the basic K$_2$HPO$_4$ to 7.8 by use of the acidic KH$_2$PO$_4$. KPi buffer and scraping buffer were stored at room temperature. Glycylglycine buffer was stored at 4°C. Aliquots were prepared of the luciferin stock and stored at -80°C. The luciferase assay mix and the luciferin solution were prepared freshly for each assay. DTT was added freshly to the KPi buffer before use.

18.b Preparation of cell extracts

For luciferase reporter-gene assays, HIT-T15 cells were cultured in 6-cm dishes and seeded at a density ~0.1 x 10^6 cells per cm². The cells were transfected with the luciferase reporter gene (3.d.II; table 3) as appropriate and additional expression plasmids depending on the experimental design. The cells were transiently transfected using DEAE-dextran (12.b). 48 h after transfection the cells were treated as indicated (13.). Afterwards, the medium was discarded and the cells were rinsed carefully with 2 mL PBS. The PBS was removed completely. The cells were detached mechanically from the plate using a scraper. They were collected in 1.5 mL scraping buffer and were transferred to a 1.5 mL tube on ice. The cells were harvested by centrifugation for 2 min at 3,000xg (Eppendorf 5417R) at 4°C. The supernatant was discarded and the pellet was resuspended in 150 µL KPi buffer. 3 cycles of freezing and thawing (liquid N_2 / 37°C waterbath) combined with intense vortexing where performed to break up the cells. The cell debris was pelleted by centrifugation for 10 min at 20,800xg (Eppendorf 5417R) at 4°C. The supernatant was subjected to determination of luciferase activity and GFPtpz fluorescence.

18.c Determination of luciferase activity

The enzyme luciferase isolated from the firefly Photinus pyralis catalyzes the adenylation of the substrate luciferin in the presence of ATP. Under O_2 consumption the luciferyl-adenylate undergoes an oxidative decarboxylation which results in the production of light. As long as the substrate luciferin is present in excess, the production of light is proportional to the amount of luciferase in the reaction mixture (de Wet et al., 1987). This allows the quantitative monitoring of promoter activity by determining the luciferase activity in luciferase reporter-gene assays.

To measure luciferase activity from cell extracts prepared from transiently transfected cells (17.b), 50 µL of the extract were mixed with 368 µL of luciferase-assay mix in a luminometer tube. The tubes were placed in the luminometer AutoLumat LB 953 (E&G Berthold). 200 µL of the luciferin solution were injected automatically to the samples and the light emission was measured at 560 nm for 20 sec. The raw data were expressed as luciferase units. Luciferin solution only was used as reference control.

18.d Measurement of GFPtpz fluorescence

Green fluorescent protein (GFP) was isolated from the jellyfish Aequorea Victoria. It contains a chromophore possessing visible absorption and fluorescence under aerobic conditions. GFP topaz (GFPtpz) is a variant of GFP which contains 4 point mutations. These point mutations modify the chromophore and thereby change the emission spectrum to yellow (Tsien, 1998). To control for transfection efficiency during luciferase reporter-gene assay the pGFPtpz-cmv® control vector was cotransfected. The expression of GFPtpz is under control of a potent CMV-promoter. Determination of GFPtpz fluorescence was performed on 96-well microplates using 50 µL of cell extracts (17.b) per sample. The fluorometer Fusion (Canberra-Packard) performed 1 sec measurement at excitation wavelength of 485 nm and emission at 530 nm. The device was operated using the software Fusion InstrumentControl Application version 3.50 (Canberra-Packard).

19. Statistics

Raw data were analyzed by one-way or two-way analysis of variance (ANOVA) followed by Student's *t*-test using the software STATISTIKA (Statsoft, Hamburg, Germany). Significance level was set at p=0.05.

Results

1. Expression and purification of GST-fusion proteins

Chemically competent *E.coli* were transformed with expression constructs coding for GST-CREB (wt or mutants), or GST alone. Bacteria colonies were screened for inducible clones, which express the GST-fusion proteins upon induction with 1 mM IPTG. Samples from bacteria colonies were analyzed by SDS-PAGE followed by Coomassie staining of the proteins. Figure 7A shows an example of the screening for inducible clones expressing GST-CREB-wt. Upon IPTG induction a band was detectable at an apparent molecular weight of ~70 kDa corresponding to the calculated molecular weight of GST-CREB-wt with 65 kDa (Figure 7A). The proteins were purified by affinity chromatography using glutathione agarose and equal amounts of proteins to be used in GST pull-down assays were determined by SDS-PAGE and Coomassie staining. Figure 7B demonstrates the comparison of GST-CREB-wt to GST-CREB-R300A and GST. For electrophoretic mobility shift assays (EMSAs) the proteins were eluted from the glutathione agarose. The concentration was determined photometrically.

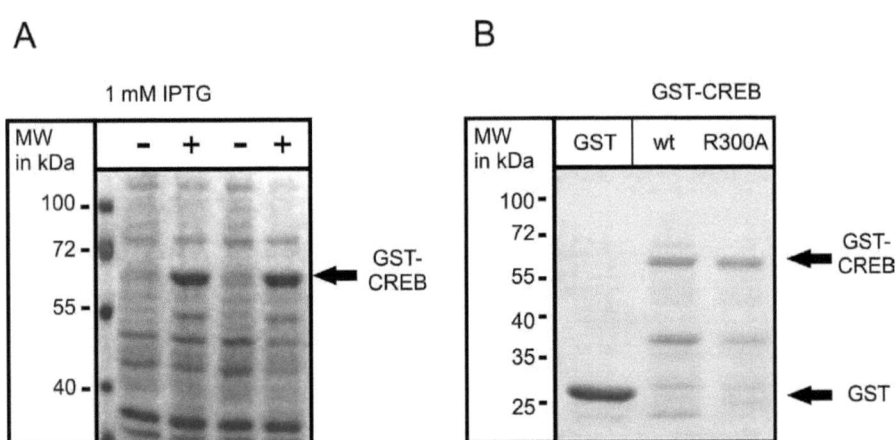

Figure 7: Results of the expression and purification of GST-fusion proteins from E.coli.

Figure 7A: Coomassie stained SDS-PAGE gel to analyze the expression of GST-CREB-wt in colonies of transformed E.coli upon induction with 1 mM IPTG. Samples of E.coli before and after 3 h of induction with IPTG are indicated with – and +, respectively. GST-CREB is detectable at an apparent molecular weight of 70 kDa upon induction with IPTG.

Figure 7B: Coomassie stained SDS-PAGE gel to analyze protein amounts and quality of purified GST-fusion proteins after purification by affinity chromatography. The GST-fusion proteins of CREB wild-type (wt) and mutant CREB-R300A (R300A) showed an apparent molecular weight of 70 kDa. GST alone showed an apparent molecular weight of 26 kDa.

2. In vitro radioactive labeling of proteins

TORC proteins were labeled radioactively by *in vitro* transcription / translation using the TNT T7 coupled reticulocyte system and [^{35}S]methionine. The coding sequences were subcloned into pcDNA3.1 under control of the T7 promoter. To control for the successful labeling the proteins were electrophoresed by SDS-PAGE and the gel was analyzed by autoradiography using a phosphor imager. Figure 8 demonstrates the results of the *in vitro* transcription / translation of [^{35}S]TORC1, [^{35}S]TORC2, and [^{35}S]TORC3. For the analysis, the [^{35}S]-labeled proteins were diluted as indicated. Bands were detectable at an apparent molecular weight of ~80 kDa, ~90 kDa, and ~75 kDa, corresponding to the calculated molecular weight of [^{35}S]TORC1 (76 kDa), [^{35}S]TORC2 (82 kDa), and [^{35}S]TORC3 (73 kDa), respectively (Figure 8).

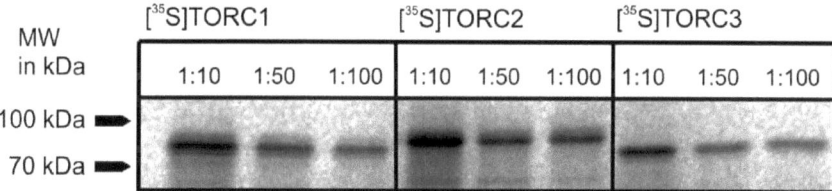

Figure 8: SDS-PAGE of [^{35}S]-labeled TORC1, TORC2, and TORC3.
After *in vitro* transcription / translation of [^{35}S]TORC1, [^{35}S]TORC2, and [^{35}S]TORC3, the proteins were electrophoresed by SDS-PAGE. The denatured proteins were loaded on the gel in a dilution of 1:10, 1:50, and 1:100. [^{35}S]TORC1 showed an apparent molecular weight of 80 kDa, whereas [^{35}S]TORC2 and [^{35}S]TORC3 showed an apparent molecular weight of 90 kDa and 75 kDa, respectively.

3. Nuclear translocation of TORC proteins in HIT-T15 cells: effect of lithium

3.a Effects of KCl and Cyclosporin A on the nuclear translocation of TORC proteins

The potential of TORC proteins to enhance CREB-directed gene transcription is tightly regulated by cytosolic and nuclear shuttling of TORC. Under resting conditions TORC is phosphorylated and sequestered in the cytoplasm bound by 14-3-3 proteins. Elevated intracellular Ca^{2+} and cAMP levels induce the dephosphorylation of TORC and its nuclear accumulation (Bittinger et al., 2004; Screaton et al., 2004). The effects of KCl and cyclosporin A (CsA) on the translocation of TORC proteins into the nucleus were analyzed by immunocytochemistry using an antibody recognizing all of the three TORC isoforms. KCl was used to depolarize the membrane of HIT-T15 cells thereby increasing intracellular Ca^{2+} levels by activation of the voltage-gated calcium channel of the L-type. Ca^{2+} activates the calcium/calmodulin-dependent phosphatase calcineurin which was shown to dephosphorylate TORC, and is potently inhibited by CsA (Bittinger et al., 2004; Screaton et al., 2004). HIT-T15 cells were treated with 45 mM KCl, 5 μM CsA, or a combination of both. Figure 9A shows typical images of cells that were untreated for control conditions (upper panel), treated with 45 mM KCl (middle panel), and with the combination of 45 mM KCl and 5 μM CsA (lower panel). For a quantitative study 150 cells per group were analyzed with respect to the localization of endogenous TORC proteins. Figure 9B shows the result of two independent experiments. The data are presented in percentage of cells with nuclear TORC. Statistical analysis by one-way ANOVA revealed specific effects of treatment with $p<0.0001$. Without treatment 16.66 ± 7.22% of counted cells showed TORC inside the nucleus. Treatment with 45 mM KCl resulted in increased amounts of cells with nuclear TORC with 81.14 ± 2.70% ($p<0.015$) compared to the control (Figure 9B). The treatment with 5 μM CsA did not affect the translocation of TORC, as in 11.52 ± 1.25% of the cells TORC was detectable in the nucleus (Figure 9B). Cotreatment of cells with KCl and CsA reduced the amount of cells with nuclear TORC to 22.65 ± 1.12%, compared to treatment with KCl alone ($p<0.0025$), to levels statistically not different from untreated cells (Figure 9B).

Results 107

A B

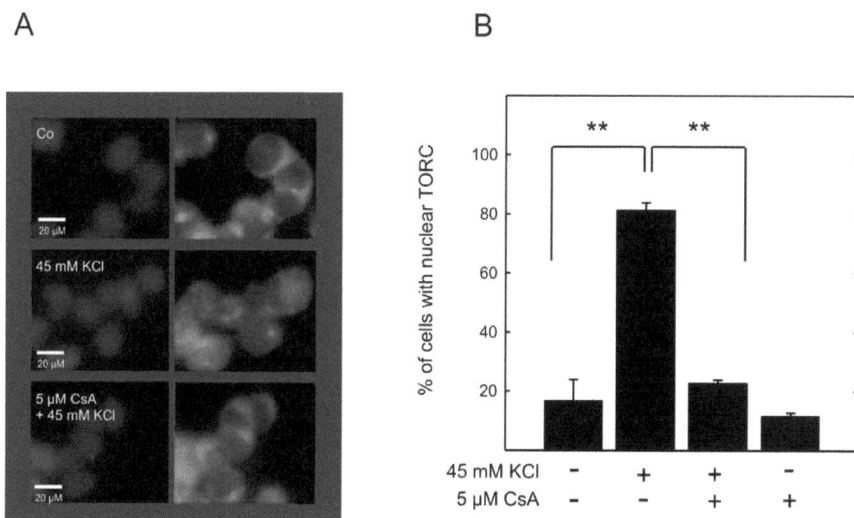

Figure 9: Nuclear translocation of endogenous TORC proteins in HIT-T15 cells upon treatment with KCl and cyclosporin A, analyzed by immunocytochemistry.
Figure 9A: Typical microscopy images of HIT-T15 cells. The nuclei were stained with DAPI, shown in blue on the left. Endogenous TORC proteins were labeled with the panTORC (1-42) antibody and the AlexaFluor®488, shown in green on the right. Co indicates the control condition without treatment on the upper panel. The middle panel shows cells treated for 30 min with 45 mM KCl. The lower panel demonstrates cells treated with the combination of 45 mM KCl and 5 μM cyclosporin A (CsA).
Figure 9B: Quantitative analysis of the nuclear translocation of endogenous TORC proteins in HIT-T15 cells. The cells were treated for 30 min with KCl and for 90 min with CsA. 150 cells were studied in each group per experiment. The values express the percentage of cells with nuclear TORC. The data are mean values ± SEM of two independent experiments. Statistical analysis was performed by one-way ANOVA, followed by Student's t-test: ** $p<0.025$. Treatment with KCl increased the percentage of cells with nuclear TORC. The treatment with CsA alone did not affect the translocation of TORC. CsA strongly reduced the nuclear accumulation of TORC induced by KCl.

3.b Effects of cAMP and lithium on the nuclear translocation of TORC proteins

The nuclear translocation of TORC proteins is also induced by elevated levels of intracellular cAMP. The salt inducible kinase (SIK) phosphorylates TORC under resting conditions. Activation of protein kinase A by cAMP reduces the activity of SIK leading to the nuclear accumulation of TORC (Takemori and Okamoto, 2008). The effects of cAMP and lithium on the nuclear translocation of TORC were investigated by immunocytochemistry as before. HIT-T15 cells were treated with 1 mM or 2 mM 8-bromo-cAMP, with 20 mM LiCl, or a combination of both. Figure 10A shows typical images of cells that were untreated (upper panel), cells that were treated with 2 mM 8-bromo-cAMP (middle panel), and cells that were treated with 20 mM LiCl (lower panel). For the quantitative analysis 150 cells were examined with respect to the localisation of endogenous TORC. The experiment was performed twice. Figure 10B shows the percentage of cells with nuclear TORC. Statistical analysis with one-way ANOVA revealed specific effects of treatment on the translocation of endogenous TORC with $p<0.0001$. Under control conditions without treatment $9.62 \pm 4.26\%$ of cells showed TORC inside the nucleus (Figure 10B). Treatment with 20 mM LiCl alone did not affect the translocation of TORC. In $13.56 \pm 4.09\%$ was TORC detectable inside the nucleus. Treatment with 1 mM or 2 mM 8-bromo-cAMP resulted in an increase of nuclear accumulation of TORC as $66.24 \pm 2.75\%$ ($p<0.011$) and $86.36 \pm 3.64\%$ ($p<0.007$) of the cells showed TORC inside of the nucleus, respectively (Figure 10B). The increase was concentration dependent as revealed by one-way ANOVA with $p<0.05$. Treatment with LiCl did not further increase the translocation induced by 1 mM or 2 mM 8-bromo-cAMP with $64.37 \pm 3.32\%$ and $93.79 \pm 1.01\%$ of cells with TORC inside the nucleus, respectively (Figure 10B).

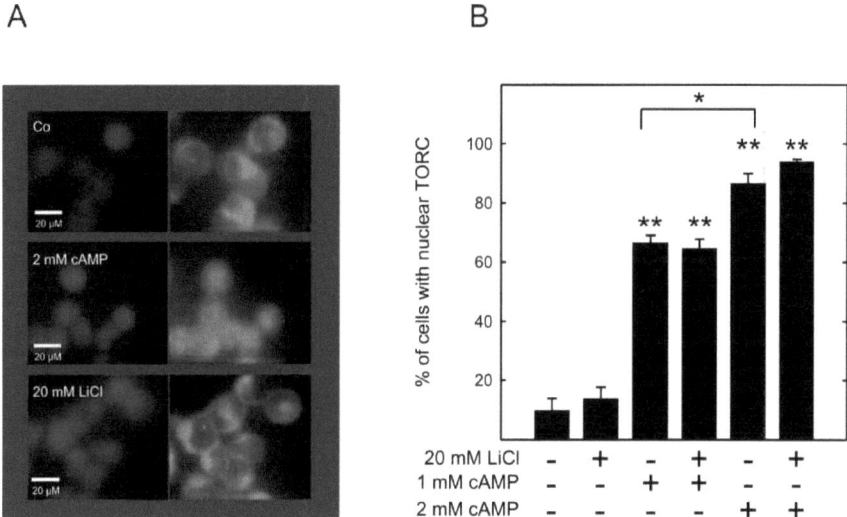

Figure 10: Nuclear translocation of endogenous TORC in HIT-T15 cells upon treatment with lithium and cAMP, analyzed by immunocytochemistry.

Figure 10A: Typical microscopy images of HIT-T15 cells. The nuclei were stained with DAPI, shown in blue on the left. Endogenous TORC proteins were labeled with the panTORC(1-42) antibody and the AlexaFluor®488, shown in green on the right. Co indicates the control condition without treatment on the upper panel. The middle panel shows cells treated for 30 min with 2 mM 8-bromo-cAMP. The lower panel demonstrates cells treated for 90 min with 20 mM LiCl.

Figure 10B: Quantitative analysis of the nuclear translocation of endogenous TORC proteins in HIT-T15 cells. The cells were treated for 90 min with LiCl and for 30 min with 8-bromo-cAMP. 150 cells were examined in each group per experiment. The values express the percentage of cells with nuclear TORC. The data are mean values ± SEM of two independent experiments. Statistical analysis was performed by one-way ANOVA, followed by Student's t-test: ** $p<0.025$; * $p<0.05$.

The treatment with LiCl alone did not affect the nuclear translocation of endogenous TORC in HIT-T15 cells. The treatment with 1 mM or 2 mM 8-bromo-cAMP increased the nuclear accumulation concentration-dependently. The translocation of TORC induced by 8-bromo-cAMP was not affected by cotreatment with LiCl.

4. Effects of lithium on the oligomerization of TORC1

TORC is able to form oligomers and was suggested to bind as a tetramer to the bZip of CREB (Conkright et al., 2003a). To examine whether lithium influences the oligomerization of TORC1 crosslinking assays were performed. On the one hand the first 44 amino acids fused to GST were used. The proteins were purified from *E.coli* by affinity chromatography and eluted from the agarose. To control for unspecific effects due to the GST-tag, GST alone was employed in the reaction. On the other hand, TORC1$_{327}$, comprising the first 327 amino acids of TORC1, was labeled radioactively by *in vitro* transcription / translation using [^{35}S]methionine. The proteins were crosslinked with glutaraldehyde with or without 20 mM LiCl and electrophoresed by SDS-PAGE. For control, 20 mM NaCl was used. Crosslinking of GST-TORC1$_{1-44}$ was analyzed by silver staining of the gel, whereas the [^{35}S]TORC1$_{327}$ was analyzed by autoradiography. Figure 11A shows a typical silver-stained gel. Under control conditions without glutaraldehyde, GST-TORC1$_{1-44}$ monomers migrate at an apparent molecular weight of 31 kDa (Figure 11A). Upon incubation of proteins with glutaraldehyde, distinct bands appeared at an apparent molecular weight of 62 kDa, 93 kDa, and 124 kDa, corresponding to dimerized, trimerized and tetramerized GST-TORC1$_{1-44}$, respectively. In the presence of 20 mM LiCl the amounts of proteins migrating at the molecular weight corresponding to dimers, trimers and tetramers of GST-TORC1$_{1-44}$ were increased (Figure 11A). This effect was not observed upon addition of 20 mM NaCl. The control GST did also demonstrate oligomerization products after crosslinking. In contrast to GST-TORC1$_{1-44}$ the amounts of GST dimers, trimers and tetramers were not increased in the presence of 20 mM LiCl (Figure 11A). To confirm these results and to avoid unspecific reactions due to the GST-tag, the crosslinking was performed with [^{35}S]TORC1$_{327}$ under similar conditions. Figure 11B shows a typical autoradiography of crosslinking of [^{35}S]TORC1$_{327}$. Monomeric [^{35}S]TORC1$_{327}$ has a calculated molecular weight of 37 kDa. To analyze putative weaker signals of the dimers, trimers and tetramers the autoradiography was performed for 30 days. Due to strong background of [^{35}S]TORC1$_{327}$ monomers, the bands corresponding to the monomers were not evaluable. Upon incubation of the proteins with glutaraldehyde, bands appeared at an apparent molecular weight of 74 kDa, 111 kDa, and 148 kDa, corresponding to the dimerized, trimerized, or tetramerized form of [^{35}S]TORC1$_{327}$, respectively (Figure 11B). 20 mM LiCl increased weakly, but detectably, the amount of protein migrating at the molecular weight corresponding to dimers, trimers, and tetramers of [^{35}S]TORC1$_{327}$ compared to the control and compared to 20 mM NaCl (Figure 11B).

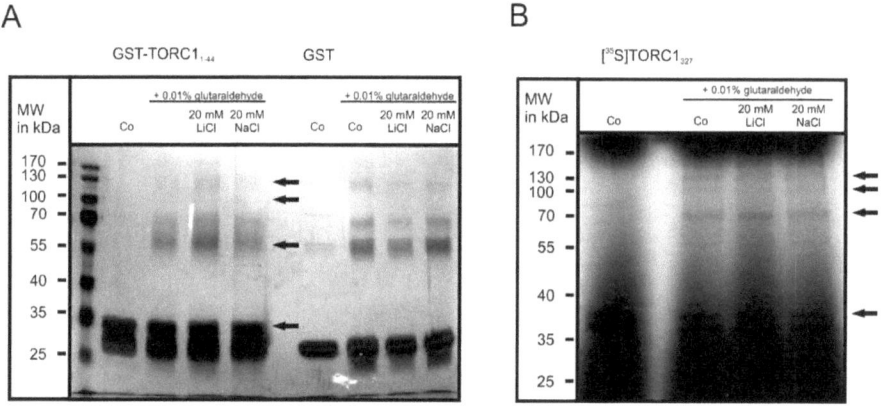

Figure 11: Effects of lithium on the oligomerization of isolated TORC1.
Figure 11A: Silver-stained SDS-PAGE of crosslinked GST-TORC1$_{1-44}$. GST-TORC1$_{1-44}$ and for control GST alone were expressed in E.coli, purified by affinity chromatography and immobilized on glutathione agarose. The proteins were eluted from the agarose and equal amounts were used per reaction. The proteins were crosslinked with 0.01% glutaraldehyde. The samples were electrophoresed by SDS-PAGE followed by silver staining of the gel. A typical gel is shown. GST-TORC1$_{1-44}$ showed an apparent molecular weight of 31 kDa. Upon treatment with glutaraldehyde dimers, trimers, and tetramers of GST-TORC1$_{1-44}$ are detectable with an apparent molecular weight of 62 kDa, 93 kDa, and 124 kDa, respectively. The presence of 20 mM LiCl increased the amount of GST-TORC1$_{1-44}$-dimers, -trimers, and tetramers, compared to the control. There was no increase detectable in the presence of 20 mM NaCl. The control GST showed an apparent molecular weight of 26 kDa. Upon incubation with glutaraldehyde, GST also showed formation of dimers, trimers, and tetramers with an apparent molecular weight of 52 kDa, 78 kDa, and 104 kDa, respectively. The presence of 20 mM LiCl or 20 mM NaCl did not increase the amounts of dimers, trimers, or tetramers compared to the control.
Figure 11B: Autoradiography of crosslinked [^{35}S]TORC1$_{327}$. TORC1$_{327}$ comprising the first 327 amino acids of TORC1 was labeled with [^{35}S]methionine by *in vitro* transcription and translation. The proteins were crosslinked with 0.01% glutaraldehyde. The samples were electrophoresed by SDS-PAGE, the gel was dried and exposed to x-ray films for 30 days. [^{35}S]TORC1$_{327}$ has a calculated molecular weight of 37 kDa. Due to a strong background signal, the bands are not distinguishable. Upon treatment with glutaraldehyde dimers, trimers, and tetramers of [^{35}S]TORC1$_{327}$ were detectable with an apparent molecular weight of 74 kDa, 111 kDa, and 148 kDa, respectively. The presence of 20 mM LiCl increased the amounts of dimers, trimers, and tetramers of [^{35}S]TORC1$_{327}$, which was not evident with 20 mM NaCl.

5. Effects of lithium on the transcriptional activity of TORC proteins

TORC proteins contain at the C-terminus a potent transactivation domain, able to induce a minimal promoter (Iourgenko et al., 2003). To investigate whether lithium influences the transcriptional activity of TORC proteins the GAL4 system was used. Here, TORC1, TORC2, or TORC3 were expressed fused to the DNA binding domain (aa 1-147) of the yeast transcription factor GAL4 (GAL4-DBD), which allows the determination of the transcriptional activity of the coactivators. Cotransfected was the luciferase-reporter gene G5E1B-Luc controlled by five repeats of the GAL4-binding site in the promoter (Figure 12, top).

Cells were treated with 1 mM 8-bromo-cAMP, 20 mM LiCl, and with the combination of both. The GAL4-DBD exhibited virtually no transcriptional activity and did not show transcriptional activation in response to treatment with 8-bromo-cAMP or LiCl (Figure 12A). The transcriptional activity of GAL4-TORC1, GAL4-TORC2, or GAL4-TORC3 was investigated in comparison with GAL4-DBD. Statistical analysis by two-way ANOVA revealed a significant difference between GAL4-DBD and GAL4 TORC-fusion proteins with $p<0.0001$. As demonstrated in Figure 12B, GAL4-TORC1, GAL4-TORC2 and GAL4-TORC3 exhibited increased basal transcriptional activity compared to GAL4-DBD. The basal activity of GAL4-TORC1 was equivalent to the one of GAL4-TORC3, but was 1.6-fold higher for GAL4-TORC2 ($p<0.006$; Figure 12B). Treatment with 20 mM LiCl did not influence the transcriptional activity of GAL4-TORC1, GAL4-TORC2, or GAL4-TORC3 (Figure 12B). 1 mM 8-bromo-cAMP increased the transcriptional activity of GAL4-TORC1 2.4-fold to 240.39 ± 20.43% ($p<0.0001$). Effects elicited by cotreatment with 8-bromo-cAMP and LiCl were equivalent to the transcriptional activity induced by 8-bromo-cAMP alone (Figure 12B). GAL4-TORC2 transcriptional activity was increased upon treatment with 1 mM 8-bromo-cAMP 1.67-fold ($p<0.002$), which was not influenced by cotreatment with 20 mM LiCl. GAL4-TORC3 did not show enhanced transcriptional activity in response to treatment with 8-bromo-cAMP alone, LiCl alone, or the combination of 8-bromo-cAMP and LiCl (Figure 12B).

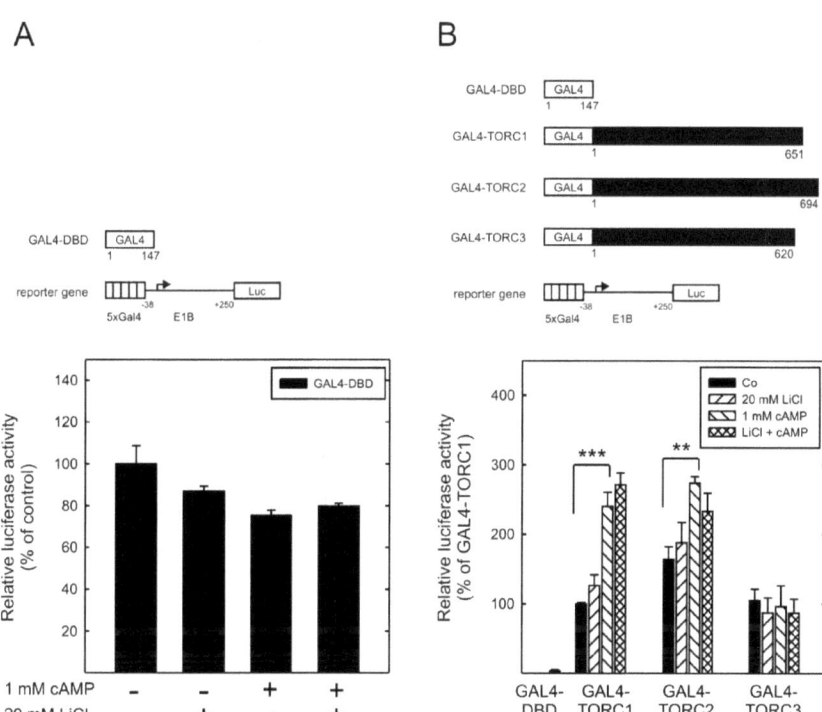

Figure 12: Effect of lithium on the transcriptional activity of TORC proteins in luciferase reporter-gene assays.

HIT-T15 cells were transiently transfected with the luciferase-reporter gene G5E1B under control of five repeats of the GAL4-binding site. Cotransfected was an expression construct encoding TORC1, TORC2, or TORC3 fused to the DNA-binding domain of GAL4, or the GAL4 DNA-binding domain (GAL4-DBD) alone. A schematic illustration of the constructs is shown above the graphs. The cells were treated with 20 mM LiCl for 7 h, with 1 mM 8-bromo-cAMP for 6 h, or with the combination of both. Luciferase activity was determined.
Figure 12A: Transcriptional activity of the GAL4 DNA-binding domain (DBD). The GAL4-DBD did not exhibit transcriptional activation upon treatment with LiCl or 8-bromo-cAMP. Relative luciferase activity values are means ± SEM of two independent experiments performed in triplicate, and are expressed in percent of control without treatment.
Figure 12B: Transcriptional activity of GAL4-TORC1, GAL4-TORC2, and GAL4-TORC3. GAL4-TORC1, GAL4-TORC2, and GAL4-TORC3 exhibited increased basal transcriptional activity compared to GAL4-DBD alone. LiCl alone had no effect on transcriptional activity. The basal activity of GAL4-TORC1 and GAL4-TORC2 was enhanced by treatment with 8-bromo-cAMP, whereas the basal activity of GAL4-TORC3 was not affected by this treatment. Cotreatment with 8-bromo-cAMP and LiCl elicited effects comparable to cAMP alone. Relative luciferase activity values are means ± SEM of three independent experiments performed in duplicate, and are expressed in percent of GAL4-TORC1 without treatment. Statistical analysis was performed by two-way ANOVA followed by Student's t-test: ***$p<0.001$, **$p<0.025$.

6. Transcriptional activity conferred by TORC1, TORC2, and TORC3 to CREB bZip – effect of lithium

It has recently been reported that TORC1 mediates the effect of lithium on cAMP-induced CREB-directed gene transcription. This was demonstrated by means of the TORC-interaction domain of CREB, bZip, expressed as GAL4-fusion protein which had no transcriptional activity on its own (Boer et al., 2007). To investigate whether all the three TORC isoforms can mediate this effect, TORC1, TORC2, and TORC3 were cotransfected with the CREB-bZip-wt fused to the DNA-binding domain of GAL4 and the luciferase reporter gene G5E1B (Figure 13, top). Cells were treated with 20 mM LiCl, with 2 mM 8-bromo-cAMP, or with the combination of both. The GAL4-bZip alone exhibited virtually no transcriptional activity and the treatment with LiCl or 8-bromo-cAMP did not result in transcriptional activation (Figure 13A). The overexpression of TORC1 resulted in increased basal transcriptional activity of the GAL4-bZip-wt ($p<0.0001$) compared to the control pBluescript (Figure 13A). Compared to TORC1, the overexpression of TORC2 increased less effectively the basal activity of GAL4-bZip-wt (24.21 ± 2.71% of TORC1, $p<0.0001$), whereas the overexpression of TORC3 enhanced the basal activity of GAL4-bZip similarly as TORC1 (Figure 13A). The treatment with 2 mM 8-bromo-cAMP increased the transcriptional activity of GAL4-bZip 3.7-fold to 371.05 ± 42.26% upon overexpression of TORC1 ($p<0.0001$). As demonstrated in Figure 13A, the treatment with 8-bromo-cAMP increased the activity of GAL4-bZip 6.2-fold to 149.51 ± 22.92% upon overexpression of TORC2 ($p<0.001$), and 3-fold to 274.59 ± 51.66% when TORC3 was overexpressed ($p<0.008$). For all TORC isoforms the treatment with LiCl increased the transcriptional activity of GAL4-bZip-wt induced by 8-bromo-cAMP, which was 3.3-fold for TORC1 ($p<0.001$), 5.3-fold for TORC2 ($p<0.005$), and 5.3-fold for TORC3 ($p<0.0001$). The analysis by two-way ANOVA confirmed significant effects for the treatment on the transcriptional activity of GAL-bZip upon overexpression of TORC proteins ($p<0.0001$), but the isoforms did not differ from each other with respect to the extent of effects elicited. Figure 13B demonstrates the effects of lithium on cAMP-induced transcriptional activity of GAL4-bZip-wt for the three different isoforms.

Results

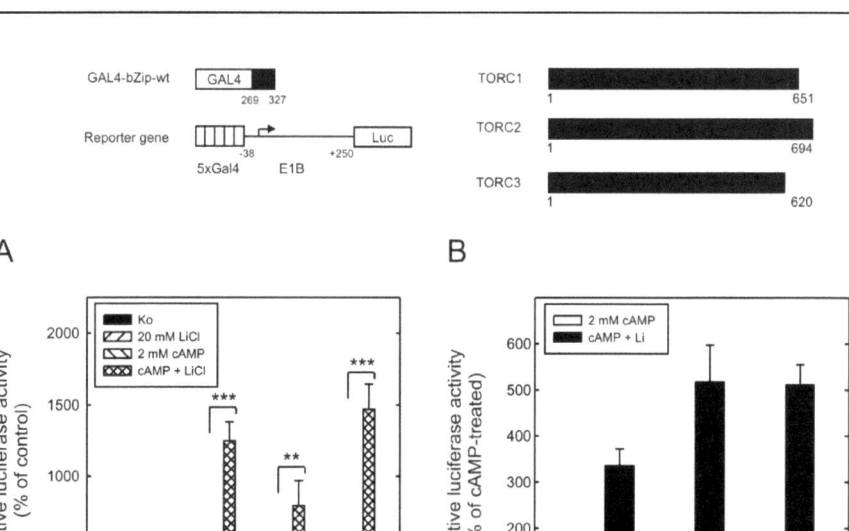

Figure 13: Effects of lithium on the transcriptional activity of GAL4-bZip upon overexpression of TORC1, TORC2, or TORC3.

HIT-T15 cells were transiently transfected with the luciferase-reporter gene G5E1B under control of five repeats of the GAL4-binding site and with an expression plasmid encoding the CREB basic leucine zipper (bZip) wild-type fused to the DNA-binding domain of GAL4. Cotransfected was an expression plasmid encoding full-length human TORC1, TORC2, or TORC3. A schematic illustration of the constructs is shown above the graphs. The cells were treated with 20 mM LiCl for 7 h, with 2 mM 8-bromo-cAMP for 6 h, or with the combination of LiCl and 8-bromo-cAMP. Luciferase activity was determined. Relative luciferase activity values are means ± SEM of three independent experiments performed in duplicate, and are expressed in percent of GAL4-bZip-wt upon overexpression of TORC1. Statistical analysis was performed by two-way ANOVA followed by Student's *t*-test: ***p<0.001, **p<0.025.

Figure 13A: Overexpression of TORC1, TORC2, or TORC3 increased the transcriptional activity of GAL4-bZip. The GAL4-bZip alone exhibited very low transcriptional activity (cotransfection of the empty vector pBluescript). Upon overexpression of TORC1, the basal activity was increased. Overexpression of TORC2, or TORC3 also resulted in increased basal activity of GAL4-bZip-wt. Compared to TORC1 the basal activity was 4-fold lower for TORC2, but similar for TORC3. The treatment with 2 mM 8-bromo-cAMP alone increased the activity of GAL4-bZip-wt upon overexpression of TORC1, TORC2, or TORC3. The enhancement by LiCl of the transcriptional activity induced by 8-bromo-cAMP was evident for all TORC isoforms.

Figure 13B: Fold induction by LiCl of cAMP-induced transcriptional activity of GAL4-bZip-wt upon overexpression of TORC1, TORC2, or TORC3. The enhancement by LiCl of the transcriptional activity of GAL4-bZip-wt induced by 8-bromo-cAMP was similar for all TORC isoforms.

7. Effects of lithium on the interaction between CREB and TORC1, TORC2, and TORC3

7.a Mammalian two-hybrid assay

To investigate effects of lithium on the interaction between CREB and TORC1, TORC2, or TORC3 in a cellular system, a mammalian two-hybrid assay was employed. Here, only the interaction domains of the proteins of interest were used. HIT-T15 cells were transiently transfected with the luciferase-reporter gene G5E1B-Luc controlled by five repeats of the GAL4-binding site and with an expression construct encoding the CREB-bZip-wt fused to the constitutively active viral protein VP16, or VP16 alone. Cotransfected was an expression plasmid coding for the first 44 amino acids of TORC1, the first 53 amino acids of TORC2, or the first 46 amino acids of TORC3 fused to the DNA-binding domain of GAL4 (Figure 14, top). Through the GAL4-DBD GAL4-TORC1$_{1-44}$, GAL4-TORC2$_{1-53}$, or GAL4-TORC3$_{1-43}$ are tethered to the promoter of the G5E1B-Luc. Transcription of the luciferase-gene is activated by VP16 if the VP16-bZip-wt interacts with a GAL4-fusion protein. Thus, luciferase activity is considered as a measure of interaction.

The results are shown in Figure 14. Compared to the control VP16, the basal activity of GAL4-TORC1$_{1-44}$ is increased 2.02-fold by VP16-bZip-wt ($p<0.0001$). VP16-bZip-wt enhanced the basal activity of GAL4-TORC2$_{1-53}$ 3.31-fold ($p<0.014$) and the basal activity of GAL4-TORC3$_{1-43}$ 2.03-fold ($p<0.048$). The treatment with 20 mM LiCl increased the luciferase activity mediated by GAL4-TORC1$_{1-44}$ 6.8-fold to 1370.17 ± 108.84% ($p<0.0001$, Figure 14). Also for GAL4-TORC2$_{1-53}$ and for GAL4-TORC3$_{1-46}$ LiCl enhanced the luciferase activity, 4.8-fold to 1596.87 ± 260.97% ($p<0.001$) and 6-fold to 1225.41 ± 187.66% ($p<0.001$), respectively (Figure 14). The statistical analysis by two-way ANOVA confirmed significant effects of lithium on the interaction between VP16-bZip-wt and GAL4-TORC1$_{1-44}$, GAL4-TORC2$_{1-53}$, or GAL4-TORC3$_{1-46}$ with $p<0.0001$, but no difference was present between GAL4-TORC1$_{1-44}$, GAL-TORC2$_{1-53}$ and GAL4-TORC3$_{1-46}$ with respect to the extent of the effects of lithium.

Results 117

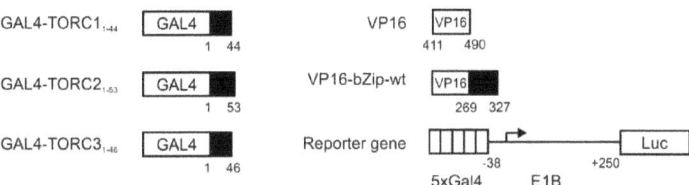

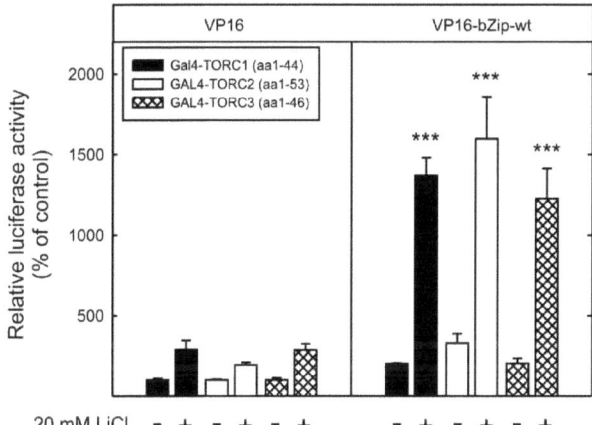

Figure 14: Effects of lithium on the interaction between CREB and TORC1, TORC2, or TORC3 in the mammalian two-hybrid assay.

HIT-T15 cells were transiently transfected with the luciferase-reporter gene G5E1B under control of five repeats of the GAL4-binding site and an expression plasmid encoding the CREB basic leucine zipper (bZip) wild-type (wt) fused to the viral protein VP16, or VP16 alone. Cotransfected was an expression construct encoding the first 44 amino acids of TORC1, the first 53 amino acids of TORC2, or the first 46 amino acids of TORC3 fused to the DNA-binding domain of GAL4. A schematic illustration of the constructs is shown above the graph. The cells were treated with 20 mM LiCl for 30 h. Relative luciferase activity values are means ± SEM of three independent experiments performed in duplicate and are expressed in percent of the control VP16 for each TORC isoform. Compared to VP16 alone, the basal activity of GAL4-TORC1$_{1-44}$ was increased upon expression of VP16-bZip-wt. 20 mM LiCl strongly increased the luciferase activity. The expression of VP16-bZip-wt also increased the basal activity of GAL4-TORC2$_{1-53}$ and GAL4-TORC3$_{1-46}$. Comparable with GAL4-TORC1$_{1-44}$ treatment with 20 mM LiCl increased the basal activity of GAL4-TORC2$_{1-53}$ and GAL4-TORC3$_{1-46}$ significantly upon expression of VP16-bZip-wt. Statistical analysis was performed by two-way ANOVA followed by Student's t-test: ***$p<0.001$.

7.b *In vitro* GST pull-down assay

7.b.I Effects of lithium on the interaction between GST-CREB and full-length [^{35}S]TORC1, [^{35}S]TORC2, or [^{35}S]TORC3

To examine the effect of lithium on the interaction between CREB and the full-length TORC1, TORC2, and TORC3 under cell-free conditions, GST pull-down assays were performed.

Bacterially expressed GST-CREB-wt immobilized on glutathione agarose (compare Figure 7B) and [^{35}S]-labeled full-length TORC1, TORC2, and TORC3 (compare Figure 8) were employed with and without LiCl in a binding reaction. The amount of [^{35}S]-labeled TORC recovered from GST-CREB-wt was determined by SDS-PAGE and subsequent autoradiography. A typical gel is shown below the graph (Figure 15). The control GST alone did not exhibit remarkable binding of [^{35}S]TORC1, [^{35}S]TORC2, or [^{35}S]TORC3 with 12.03 ± 4.02% ($p<0.001$), 2.81 ± 0.40% ($p<0.0001$), and 20.53 ± 7.31% ($p<0.002$) of GST-CREB-wt, respectively (Figure 15), which was not changed by the presence of 20 mM LiCl. For GST-CREB-wt 20 mM LiCl increased the amount of [^{35}S]TORC1 recovered from the sample to 175.84 ± 7.96% ($p<0.006$). In contrast, the amount of [^{35}S]TORC2 or [^{35}S]TORC3 recovered from GST-CREB-wt was not changed (Figure 15), as confirmed by two-way ANOVA.

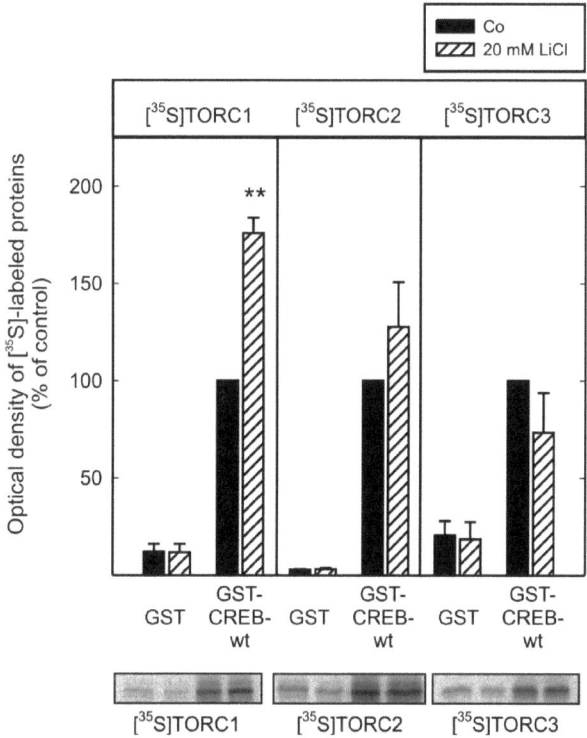

Figure 15: Effects of lithium on the interaction between CREB and full-length TORC1, TORC2, or TORC3 in the GST pull-down assay.

GST-CREB wild-type (wt) and for control GST alone were expressed in E.coli, purified by affinity chromatography and immobilized on glutathione agarose. TORC1, TORC2, or TORC3 were labeled with [^{35}S]methionine by in vitro transcription and translation. The amounts of [^{35}S]-labeled proteins recovered from GST-CREB-wt or GST were analyzed by SDS-PAGE followed by autoradiography and densitometric analysis of the bands corresponding to [^{35}S]TORC. Typical gels are shown below the graph. The data are mean values ± SEM of four independent experiments and are expressed in percent of GST-CREB-wt per group. Statistical analysis was performed by two-way ANOVA followed by two-sided paired Student's t-test: **p<0.025.
GST alone exhibited no remarkable binding of [^{35}S]TORC1, [^{35}S]TORC2, or [^{35}S]TORC3, compared to GST-CREB-wt. 20 mM LiCl increased the amount of [^{35}S]TORC1 recovered from GST-CREB-wt. The binding of [^{35}S]TORC2 or [^{35}S]TORC3 was not affected by the presence of 20 mM LiCl.

7.b.II Effects of lithium on the interaction between GST-CREB-wt and truncated [^{35}S]TORC1$_{327}$, [^{35}S]TORC2$_{347}$, or [^{35}S]TORC3$_{310}$

To examine a putative difference between full-length TORC proteins and shorter fragments for the effect of lithium on their interaction with GST-CREB-wt, truncated TORC proteins were employed in GST pull-down assays. Bacterially expressed GST-CREB-wt immobilized on glutathione agarose (compare Figure 7B) and [^{35}S]-labeled TORC proteins comprising the first 327, 347, or 310 amino acids of TORC1, TORC2, or TORC3, respectively, were employed with and without LiCl in a binding reaction. The samples were analyzed by SDS-PAGE and autoradiography followed by densitometric analysis of the bands corresponding to the [^{35}S]-labeled proteins. A typical gel is shown below the graph (Figure 16). The negative control GST alone did not exhibit remarkable binding of [^{35}S]TORC1$_{327}$, [^{35}S]TORC2$_{347}$, or [^{35}S]TORC3$_{310}$ with 5.07 ± 1.98% ($p<0.0001$), 1.85 ± 0.41% ($p<0.0001$), and 8.94 ± 2.70% ($p<0.0001$) of GST-CREB-wt, respectively (Figure 16), which was not altered by the presence of 20 mM LiCl. The amount of [^{35}S]TORC1$_{327}$ recovered from the GST-CREB-wt was increased by lithium to 213.19 ± 45.48% ($p<0.05$). The presence of 20 mM LiCl did not change the amount of [^{35}S]TORC2$_{347}$ or [^{35}S]TORC3$_{310}$ recovered from GST-CREB-wt (Figure 16), as confirmed by two-way ANOVA.

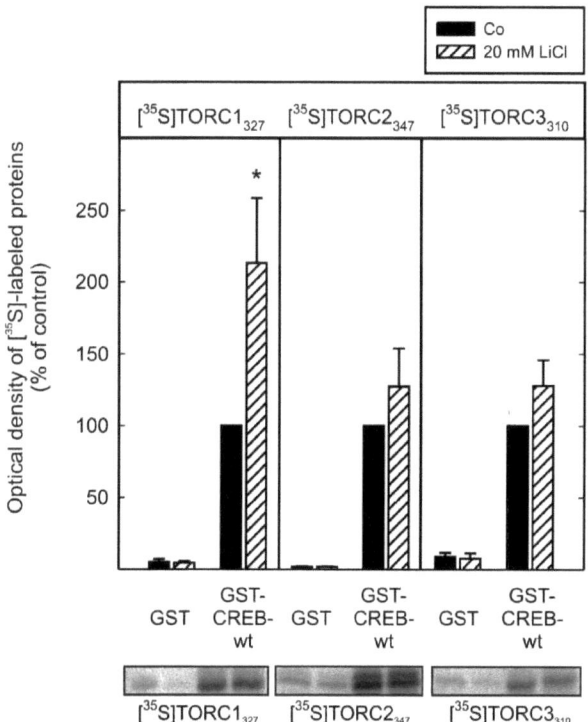

Figure 16: Effects of lithium on the interaction between CREB and truncated TORC1$_{327}$, TORC2$_{347}$, or TORC3$_{310}$ in the GST pull-down assay.
GST-CREB wild-type (wt) and, for control, GST alone were expressed in E.coli, purified by affinity chromatography and immobilized on glutathione agarose. TORC1$_{327}$, TORC2$_{347}$, or TORC3$_{310}$ comprising the first 327, 347, or 310 amino acids of TORC1, TORC2, or TORC3, respectively, were labeled with [^{35}S] by in vitro transcription and translation. The amounts of [^{35}S]-labeled proteins recovered from GST-CREB-wt or GST were analyzed by SDS-PAGE followed by autoradiography and densitometric analysis of the bands corresponding to [^{35}S]TORC. Typical gels are shown below the graph. The data are mean values ± SEM of four independent experiments and are expressed in percent on GST-CREB-wt per group. Statistical analysis was performed by two-way ANOVA followed by two-sided paired t-test: *p<0.05.
GST alone exhibited no remarkable binding of [^{35}S]TORC1$_{327}$, [^{35}S]TORC2$_{347}$, or [^{35}S]TORC3$_{310}$, compared to GST-CREB-wt. 20 mM LiCl increased the amount of [^{35}S]TORC1$_{327}$ recovered from GST-CREB-wt. The binding of [^{35}S]TORC2$_{347}$ or [^{35}S]TORC3$_{310}$ was not affected by the presence of 20 mM LiCl.

8. Identification of TORC isoforms expressed in HIT-T15 cells

In the present study, the electrically excitable cell line HIT-T15 was used to investigate the enhancement by lithium of the cAMP-induced CREB-directed gene transcription. Here, the expression levels of the different TORC isoforms in HIT-T15 were analyzed. For this purpose, total RNA was extracted, cDNA was generated by reverse transcription using an oligo-dT primer, and the amounts of TORC1, TORC2, and TORC3 transcripts were quantified by PCR using specific primers (Figure 17A).

Figure 17B shows the result of the quantitative analysis of the bands by densitometry. The experiments revealed the strong expression of TORC1 in HIT-T15 cells, whereas TORC2 expression was 8-fold lower with 12.43 ± 0.02% compared to TORC1. TORC3 transcripts were not detectable in HIT-T15 cells. The cDNA of the hamster TORC1 isoform extracted from HIT-T15 cells was subcloned into pcDNA3.1 and sequenced (Appendix A). The first 243 bases of hamster TORC1 showed 93% homology to human TORC1, whereas bases 244 to 1793 exhibited 84% homology to human TORC1.

Due to the result that TORC1 is the isoform mainly expressed in HIT-T15 cells the following experiments issued only TORC1.

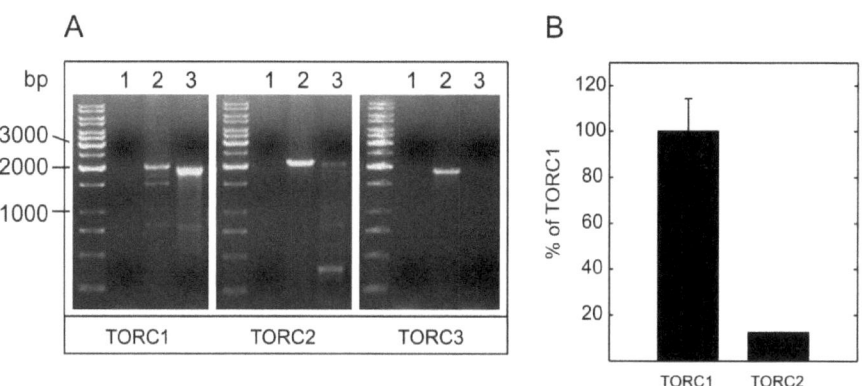

Figure 17: Expression levels of TORC isoforms in HIT-T15 cells.
Total RNA was extracted from HIT-T15 cells. cDNA was generated from mRNA by use of oligo(dT)$_{15}$ primer and M-MLV Reverse Transcriptase. The level of TORC mRNA present in HIT-T15 cells was analyzed by PCR using TORC1-, TORC2-, and TORC3-specific primers.
Figure 17A: Agarose gel of the PCR samples from cDNA generated from HIT-T15 cells. After PCR the samples were electrophoresed on an agarose gel. Lane 1 indicates the PCR negative control without template. Lane 2 indicates the PCR positive control using the human TORC1, TORC2, or TORC3 cDNA as template. Lane 3 indicates the result of the PCR from the cDNA generated from mRNA of HIT-T15 cells.
Figure 17B: Densitometric analysis of the PCR samples. TORC1 is the isoform expressed at highest level. TORC2 is expressed 8-fold lower compared to TORC1, and TORC3 is not expressed. The data are mean values ± SEM of two independent experiments performed in duplicate and are expressed in percent of TORC1 mRNA level.

9. Examination of the interaction between CREB and TORC1 at the somatostatin CRE in the EMSA

To demonstrate the interaction between CREB and TORC1 at the CRE of the rat *somatostatin*-gene promoter, electrophoretic mobility shift assays (EMSAs) were performed. Bacterially expressed GST-CREB-wt (compare Figure 7B) and *in vitro* transcribed / translated TORC1 were employed together with the [^{32}P]-labeled double-stranded oligomer containing the rat *somatostatin* promoter CRE. Figure 18 shows a typical gel. Compared to the free probe (Figure 18, lane 10) a clear band shift is detectable when the probe [^{32}P]somCRE was incubated with GST-CREB. The addition of TORC1 and GST-CREB-wt slightly shifts the band compared to GST-CREB-wt alone (Figure 18, lane 2). Noteworthy, the addition of TORC1 alone also lead to a bandshift (Figure 18, lane 9). Since TORC1 was not purified from the reticulocyte lysate of the transcription / translation kit, other proteins present in the lysate might bind to the probe. In order to examine whether TORC and CREB form a complex on the probe, a supershift was performed. The preincubation of the proteins with CREB-KID antibody led to a slight, but detectable, band shift of the complex depending on the amount of antibody (Figure 18, lanes 3, 4, 5). Also, the preincubation of the proteins with the panTORC antibody resulted in a band shift of the complex (Figure 18, lane 6, 7, 8) compared to the sample without antibody (Figure 18, lane 2).

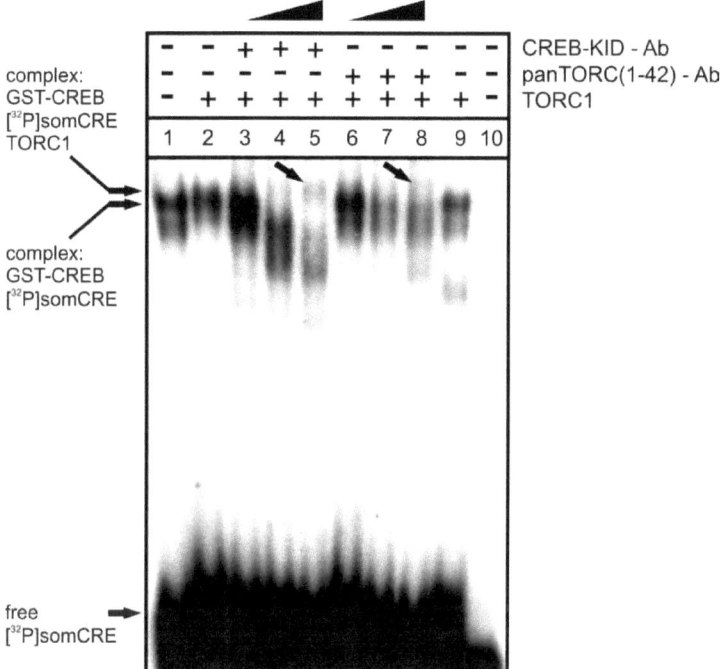

Figure 18: CREB and TORC1 interaction at the *somatostatin* CRE in electrophoretic mobility shift assays.

A typical gel of electrophoretic mobility shift assays (EMSAs) is shown.
200 ng of bacterially expressed GST-CREB-wt were used per reaction with [^{32}P]somCRE, containing the CRE of the rat *somatostatin*-gene promoter (somCRE). 7 µl of *in vitro* transcribed and translated TORC1 were supplemented as indicated above the gel. Increasing amounts of either CREB-KID antibody or panTORC (1-42) antibody are marked at the top.
Lane 1: GST-CREB-wt and [^{32}P]somCRE;
Lane 2: GST-CREB-wt, TORC1, and [^{32}P]somCRE;
Lane 3: GST-CREB-wt, TORC1, and [^{32}P]somCRE + 1 µl anti-CREB-KID;
Lane 4: GST-CREB-wt, TORC1, and [^{32}P]somCRE + 3 µl anti-CREB-KID;
Lane 5: GST-CREB-wt, TORC1, and [^{32}P]somCRE + 6 µl anti-CREB-KID;
Lane 6: GST-CREB-wt, TORC1, and [^{32}P]somCRE + 1 µl anti-panTORC;
Lane 7: GST-CREB-wt, TORC1, and [^{32}P]somCRE + 3 µl anti-panTORC;
Lane 8: GST-CREB-wt, TORC1, and [^{32}P]somCRE + 6 µl anti-panTORC;
Lane 9: TORC1 and [^{32}P]somCRE;
Lane 10: free [^{32}P]somCRE.
GST-CREB-wt alone or in combination with TORC1 led to a band shift of the [^{32}P]somCRE. A slight band shift of the complex was detectable upon addition of 6 µl of either anti-CREB-KID antibody or anti-panTORC antibody, as marked by arrows on the gel.

10. Effects of lithium on the recruitment of TORC1 to the promoter

To investigate effects of lithium and cAMP on the recruitment of TORC1 to the promoter chromatin-immunoprecipitation (ChIP) assays were performed. HIT-T15 cells were transiently transfected with an expression construct coding for TORC1 tagged with a FLAG-epitope. Cotransfected was 4xsomCRE-Luc containing four repeats of the CRE-consensus sequence 5'-TGACGTCA-3' from the rat *somatostatin* gene promoter as a binding site for endogenous CREB (Figure 19, top). Cells were treated with 20 mM LiCl, with 2 mM 8-bromo-cAMP, and with a combination of both. Subsequently, proteins and DNA were crosslinked and precipitated with the FLAG M2 antibody. The amount of precipitated promoter DNA was quantified by real-time PCR. The data given in Figure 19 represent the amount of precipitated 4xsomCRE-DNA molecules expressed in percent of control. The control pBluescript did virtually exhibit no specific signal in the real-time PCR (Figure 19A). Treatment of HIT-T15 cell with 20 mM LiCl alone or 2 mM 8-bromo-cAMP alone did not induce a detectable increase in promoter occupancy by FLAG-TORC1 with 128.66 ± 84.65% and 90.22 ± 32.41% of control. Treatment with the combination of LiCl and 8-bromo-cAMP increased the promoter occupancy by FLAG-TORC1 about 2-fold with 187.69 ± 55.94% (Figure 19B). This increase is statistically significant with $p<0.04$ compared to untreated cells and $p<0.025$ compared to treatment with 8-bromo-cAMP alone (Figure 19B).

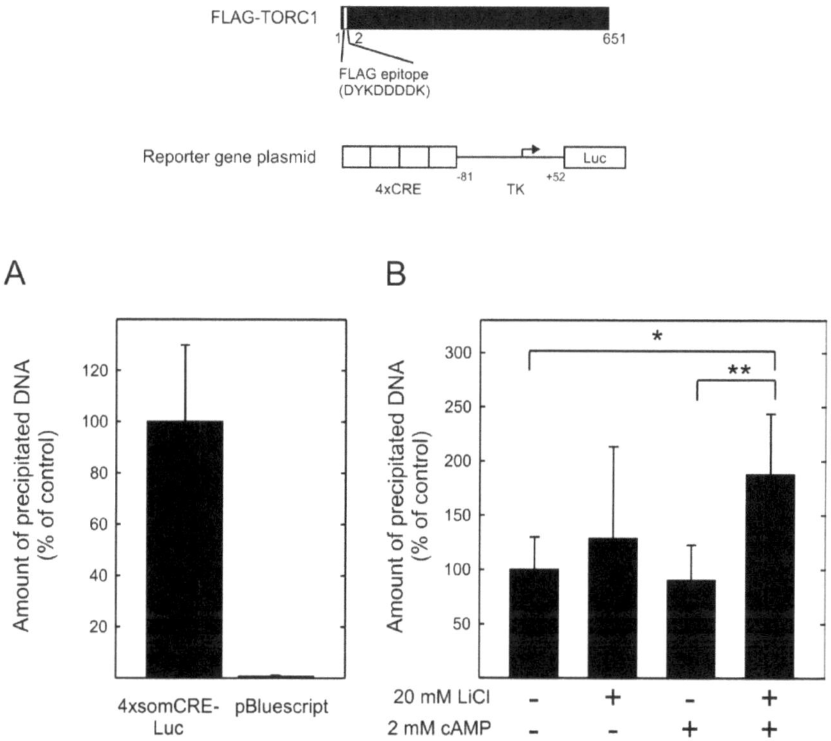

Figure 19: Effects of lithium and cAMP on the recruitment of TORC1 to promoter in chromatin immunoprecipitation assays.

For chromatin immunoprecipitation (ChIP) assays, HIT-T15 cells were transiently transfected with the reporter gene plasmid 4xsomCRE containing 4 repeats of the CRE of the rat *somatostatin*-gene promoter in front of the truncated thymidine kinase promoter (TK). Cotransfected was an expression plasmid encoding the FLAG-tagged TORC1. A schematic illustration of the constructs is given above the graphs. For control conditions the empty vector pBluescript was cotransfected. The cells were treated for 30 min with 2 mM 8-bromo-cAMP, for 90 min with LiCl, or with the combination of 8-bromo-cAMP and LiCl. DNA and proteins were crosslinked with 1% formaldehyde, the complexes were precipitated by use of the FLAG M2 antibody, and the amount of precipitated promoter DNA was determined by quantitative real-time PCR. Mean values ± SEM of 5 independent experiments are shown. Values are the amount of precipitated DNA molecules expressed in percent of control.

Figure 19A: Specificity control of the assay. The control exhibited no specific signal in quantitative real-time PCR.

Figure 19B: Effects of treatment of the recruitment of TORC1 to the promoter 4xsomCRE. The treatment with LiCl or 8-bromo-cAMP alone did not result in a detectable increase in promoter occupancy by FLAG-TORC1. LiCl combined with 8-bromo-cAMP increased the amount of FLAG-TORC1 recruited to the 4xsomCRE promoter. Statistical analysis was performed by two-way ANOVA followed by Student's *t*-test: **$p<0.025$; *$p<0.05$.

11. Further characterization of the interaction between CREB and TORC1

11.a Requirement of the first 44 amino acids of TORC1 to confer transcriptional activity to CREB bZip

A TORC1 construct lacking the first 44 amino acids was employed. HIT-T15 cells were transiently transfected with the luciferase-reporter gene G5E1B-Luc controlled by 5 repeats of the GAL4-binding site and with an expression plasmid encoding the GAL4-bZip. Cotransfected was an expression plasmid encoding either the full-length human TORC1, or TORC1$_{\Delta 44}$ comprising amino acids 45-651 of TORC1 (Figure 20, left). The cells were treated with 20 mM LiCl, with 1 mM 8-bromo-cAMP, or with the combination of both. As shown above, the overexpression of full-length TORC1 increased the basal transcriptional activity of GAL4-bZip-wt and treatment with lithium significantly enhanced the cAMP-induced activity 2.72-fold (p<0.004; Figure 20). In contrast, the overexpression of TORC1$_{\Delta 44}$ did not confer activity to GAL4-bZip-wt, neither did the empty vector pBluescript (Figure 20) as confirmed by two-way ANOVA with p<0.0001. These results underline the relevance of the CREB-interaction domain of TORC1 for the observed effects.

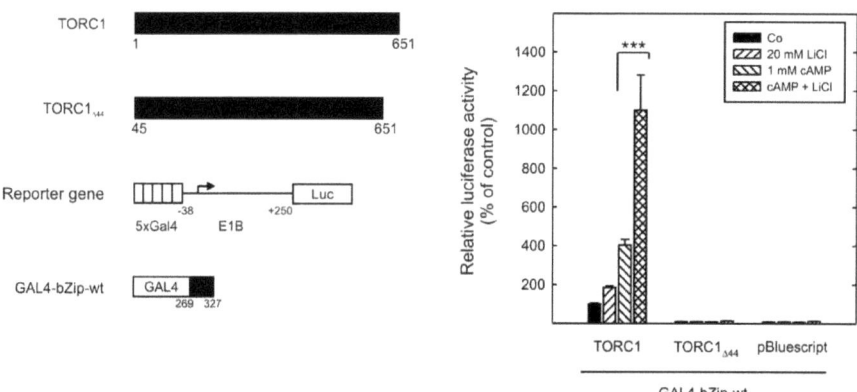

Figure 20: Requirement of the first 44 amino acids of TORC1 to confer transcriptional activity to GAL4-bZip-wt in luciferase reporter-gene assays.

HIT-T15 cells were transiently transfected with the luciferase-reporter gene G5E1B under control of five repeats of the GAL4-binding site and with an expression plasmid encoding the CREB-bZip wild-type (wt) fused to the DNA-binding domain of GAL4. Cotransfected was an expression plasmid encoding full-length human TORC1, or TORC1 truncated by the first 44 amino acids (TORC$_{\Delta 44}$). For control the empty vector pBluescript was cotransfected. A schematic illustration of the constructs is shown. The cells were treated with 20 mM LiCl for 7 h, with 1 mM 8-bromo-cAMP for 6 h, or with the combination of both. Luciferase activity was determined. Relative luciferase activity values are means ± SEM of three independent experiments performed in duplicate, and are expressed in percent of GAL4-bZip-wt upon overexpression of TORC1. The full-length TORC1 mediates transcriptional activation of the GAL4-bZip-wt. Treatment with LiCl alone or 8-bromo-cAMP alone increased the transcriptional activity. LiCl enhanced the activity induced by 8-bromo-cAMP. Overexpression of TORC$_{\Delta 44}$ did not result in transcriptional activation of GAL4-bZip-wt as comparable to the control pBluescript. Statistical analysis was performed by two-way ANOVA followed by Student's t-test: ***$p<0.001$.

11.b Effect of the R300A mutation of CREB bZip on the interaction between CREB and TORC1 as revealed by the mammalian two-hybrid assay

In addition to CREB wild-type the mutant R300A was used, which has been reported to be TORC-binding deficient (Screaton et al., 2004). HIT-T15 cells were transfected with the luciferase-reporter gene G5E1B-Luc coding for the luciferase gene under control by five repeats of the GAL4-binding site and an expression vector coding for the first 44 amino acids of TORC1 fused to the DNA-binding domain of the yeast transcription factor GAL4. Cotransfected was an expression construct coding either for the CREB basic leucine zipper (bZip) wild-type (wt), or the mutant bZip-R300A, fused to the constitutively active viral protein VP16. To control for the specificity of the interaction an expression vector encoding VP16 alone was cotransfected (Figure 21, top). The cells were treated with 20 mM LiCl, with 1 mM 8-bromo-cAMP, or with the combination of both, and luciferase activity was determined. The data given in Figure 21 were kindly provided by Ulrike Böer. As shown above (Figure 14), expression of VP16-bZip-wt increased 2-fold the basal activity of GAL4-TORC1$_{1-44}$ compared to VP16 alone (243.07 ± 13.65%, $p<0.0001$). For the VP16-bZip-wt the treatment with 20 mM LiCl increased the luciferase activity 6-fold to 1489.79 ± 124.14% ($p<0.0001$; Figure 21). 8-bromo-cAMP did not affect luciferase activity. The cotreatment of 8-bromo-cAMP with LiCl resulted in values similar to LiCl alone (Figure 21). The mutant VP16-bZip-R300A did not show significant differences from the control VP16 alone (Figure 21). Statistical analysis by two-way ANOVA confirmed significant effects of treatment with LiCl with $p<0.0001$ and significant differences between the VP16-bZip-wt and VP16 or VP16-bZip-R300A, respectively, with $p<0.0001$. Effects of treatment were different depending on the type of protein with $p<0.0001$.

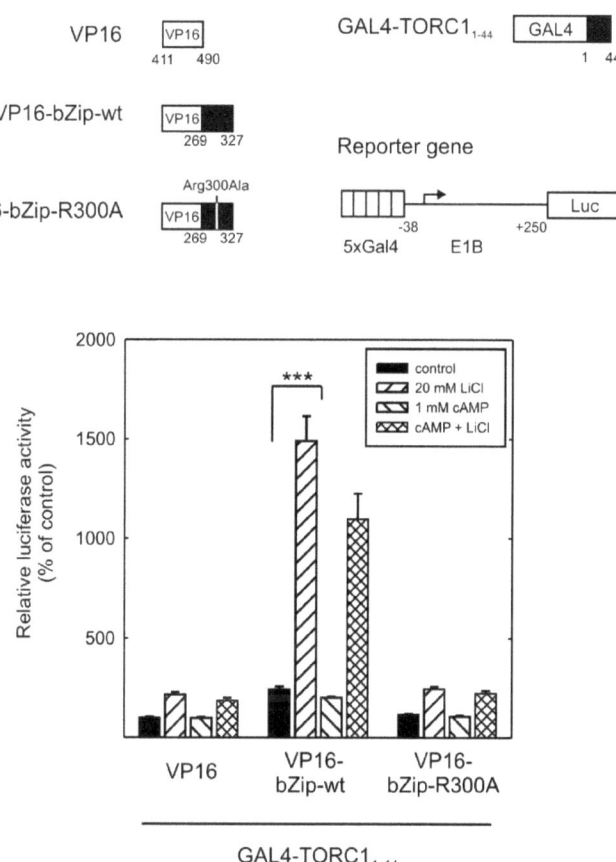

Figure 21: Effects of lithium and cAMP on the interaction between CREB or CREB-R300A and TORC1 in the mammalian two-hybrid assay.

HIT-T15 cells were transiently transfected with the luciferase-reporter gene G5E1B under control of five repeats of the GAL4-binding site and an expression construct encoding the first 44 amino acids of TORC1 fused to the DNA-binding domain of GAL4. Cotransfected was an expression plasmid encoding the CREB basic leucine zipper (bZip) wild-type (wt) or mutant R300A fused to the viral protein VP16, or VP16 alone. A schematic illustration of the constructs is shown. The cells were treated with 20 mM LiCl for 30 h, with 1 mM 8-bromo-cAMP for 6 h, or with the combination of 20 mM LiCl and 1 mM 8-bromo-cAMP. Relative luciferase activity values are means ± SEM of three independent experiments performed in duplicate and are expressed in percent of the control VP16. Compared to VP16 alone, the basal activity of GAL4-TORC1$_{1-44}$ was increased upon expression of VP16-bZip-wt. 20 mM LiCl increased the luciferase activity 6-fold. 1 mM 8-bromo-cAMP did not affect the transcriptional activity. VP16-bZip-R300A was not different from the control VP16 alone. Statistical analysis was performed by two-way ANOVA followed by Student's t-test: ***$p<0.001$. The data presented here were kindly provided by Ulrike Böer.

11.c In vitro GST pull-down assay

11.c.I Concentration-response curve of the effect of lithium on the specific interaction between GST-CREB and [^{35}S]TORC1

Bacterially expressed GST-CREB-wt and GST-CREB-R300A immobilized on glutathione agarose (compare Figure 7B) and [^{35}S]-labeled full-length TORC1 (compare Figure 8) were employed with and without LiCl in a binding reaction. The amount of [^{35}S]-labeled TORC1 recovered from GST-fusion proteins was determined by SDS-PAGE and subsequent autoradiography. Typical gels are pictured below the graphs (Figure 22). The control GST alone exhibited no remarkable recovery of [^{35}S]TORC1 with 6.6 ± 3.08% compared to GST-CREB-wt (p<0.002), which was not changed by LiCl (Figure 22A). In the presence of 20 mM LiCl, the amount of [^{35}S]TORC1 recovered from GST-CREB-wt was increased to 155.72 ± 10.13% (p<0.035) (Figure 22A), as was shown above (compare Figure 15). For the mutant GST-CREB-R300A no specific binding of [^{35}S]TORC1 was detectable with 5.07 ± 3.23% compared to GST-CREB-wt. The presence of LiCl did not alter the binding of [^{35}S]TORC1 to GST-CREB-R300A (Figure 22A).

Figure 22B demonstrates a concentration-response curve of LiCl. Regarding the interaction between CREB and TORC1, the concentration-response curve revealed significantly increased amounts of [^{35}S]TORC1 recovered from GST-CREB-wt at concentrations of 5 mM LiCl with 157.68 ± 21.19% compared to the control (p<0.035). At 20 mM LiCl 191.63 ± 18.51% of [^{35}S]TORC1 were recovered from GST-CREB-wt (p<0.003) compared to the control (Figure 22B). Data analysis by one-way ANOVA resulted in a significant effect of the presence of lithium on the binding of [^{35}S]TORC1 to GST-CREB-wt with p<0.013.

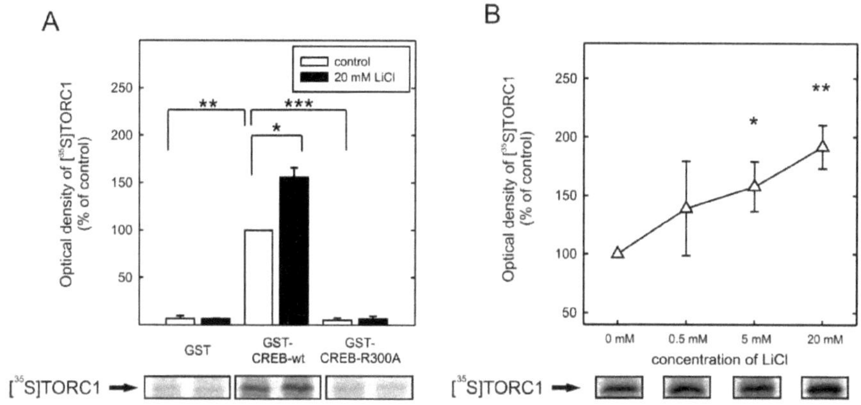

Figure 22: Concentration-response curve of the effect of lithium on the specific interaction between GST-CREB and [^{35}S]TORC1 in a GST pull-down assay.

GST-CREB wild-type (wt) and mutant R300A, and for control GST alone, were expressed in *E.coli*, purified by affinity chromatography and immobilized on glutathione agarose. TORC1 was labeled with [^{35}S]methionine by *in vitro* transcription and translation. The amount of [^{35}S]TORC1 recovered from GST-fusion proteins was analyzed by SDS-PAGE followed by autoradiography and densitometric analysis of the bands corresponding to [^{35}S]TORC1. Typical gels are shown below the graphs.

Figure 22A: Lithium increased the interaction between GST-CREB and [^{35}S]TORC1 in a mutation-sensitve manner. GST alone exhibited no remarkable binding of [^{35}S]TORC1 compared to GST-CREB-wt. 20 mM LiCl increased the amount of [^{35}S]TORC1 recovered from GST-CREB-wt. The mutant GST-CREB-R300A was not different from the control GST alone. The data are mean values ± SEM of three independent experiments and are expressed in percent of GST-CREB-wt.

Figure 22B: Lithium enhanced the interaction between GST-CREB and [^{35}S]TORC1 in a concentration-dependent manner. The dose-response curve revealed significantly increased amounts of [^{35}S]TORC1 recovered from GST-CREB-wt at concentrations of 5 mM LiCl and 20 mM LiCl. The data are mean values ± SEM of seven independent experiments and are expressed in percent of GST-CREB-wt.

Statistical analysis was performed by two-way ANOVA followed by two-sided paired Student's *t*-test: ***$p<0.001$; **$p<0.025$; *$p<0.05$.

11.c.II Effect of magnesium on the interaction between GST-CREB and [^{35}S]TORC1

The lithium ion has a similar ionic radius compared with the magnesium ion and shares with magnesium ligand binding properties. Moreover lithium was shown to compete with magnesium for magnesium binding sites in enzymes (Mota de Freitas et al., 2006). To examine the effect of magnesium ions on the interaction between CREB and TORC1 GST pull-down assays with recombinant GST-CREB-wt and [^{35}S]TORC1 in the presence of increasing amounts of magnesium were performed.

Figure 23A shows the concentration-response curve of increasing amounts of magnesium, demonstrating a strong reduction of the amounts of [^{35}S]TORC1 recovered from GST-CREB-wt. At 0.5 mM MgCl$_2$ the amount of [^{35}S]TORC1 bound by GST-CREB-wt was reduced to 59.92 ± 6.79% compared to the control without MgCl$_2$ ($p<0.028$). At 5 mM MgCl$_2$ and at 10 mM MgCl$_2$ the amount of recovered [^{35}S]TORC1 was reduced to 36.80 ± 5.81% ($p<0.009$) and 24.06 ± 1.44% ($p<0.001$), respectively. At 20 mM MgCl$_2$ the binding of [^{35}S]TORC1 was reduced to 14.29 ± 4.11% ($p<0.003$) which was statistically not different from the unspecific control GST alone. The data analysis by one-way ANOVA confirmed the effect of magnesium on the binding of [^{35}S]TORC1 to GST-CREB-wt with $p<0.0001$. Next, the influence of lithium on this inhibitory effect of magnesium was examined. In the presence of 5 mM LiCl the strong inhibition of the interaction between GST-CREB-wt and [^{35}S]TORC1 by MgCl$_2$ was attenuated as shown in Firgure 23B and Figure 23C. 5 mM LiCl significantly increased the interaction between GST-CREB-wt and [^{35}S]TORC1 in presence of 10 mM or 20 mM MgCl$_2$ by 270.16 ± 53.42% or 280.18 ± 52.68% ($p<0.044$ and $p<0.04$), respectively. Statistical analysis by two-way ANOVA resulted in a significant effect of LiCl depending on the concentration of MgCl$_2$ with $p<0.046$.

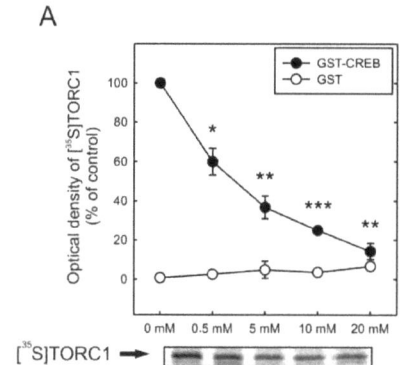
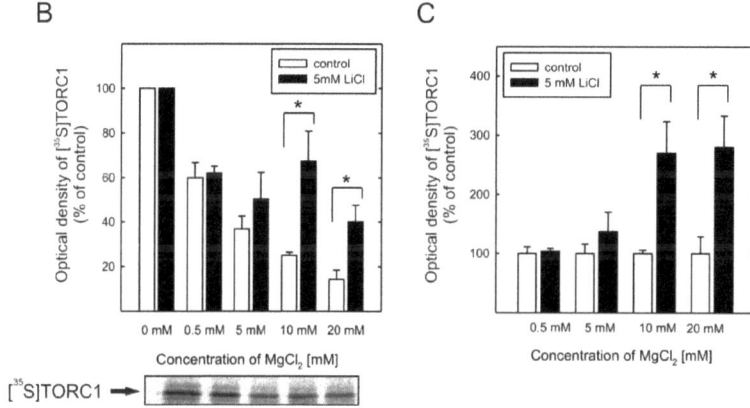

Figure 23: Effect of magnesium on the interaction between CREB and TORC1 in the GST pull-down assay.

GST-CREB wild-type (wt) and, for control, GST alone, were expressed in E.coli, purified by affinity chromatography and immobilized on glutathione agarose. TORC1 was labeled with [^{35}S] by *in vitro* transcription and translation. The amount of [^{35}S]TORC1 recovered from GST-fusion proteins was analyzed by SDS-PAGE followed by autoradiography and densitometric analysis of the bands corresponding to [^{35}S]TORC1. Typical gels are shown below the graphs. The data are mean values ± SEM of three independent experiments and are expressed in percent of GST-CREB-wt. Statistical analysis was performed by two-way ANOVA followed by two-sided paired Student's *t*-test: ***$p<0.001$; **$p<0.025$; *$p<0.05$.

Figure 23A: Magnesium inhibits the interaction between GST-CREB and [^{35}S]TORC1 in a concentration-dependent manner. The dose-response curve reveals significantly reduced amounts of [^{35}S]TORC1 recovered from GST-CREB-wt at concentrations of 0.5 mM MgCl$_2$.

Figure 23B: 5 mM LiCl reduced the strong inhibition of the interaction between GST-CREB-wt and [^{35}S]TORC1 at 10 mM and 20 mM MgCl$_2$.

Figure 23C: To compare the effects with and without 5 mM LiCl, the data without LiCl were set as 100% in each group. 5 mM LiCl enhanced the interaction 3-fold at 10 mM and 20 mM MgCl$_2$.

12. Mutation of CREB at Lysine 290 (K290)

The X-ray structure of the bZip of CREB bound to the CRE of the rat *somatostatin* promoter revealed the presence of a hexahydrated magnesium ion. This magnesium ion is coordinated by the lysine residues at position 290 (K290) of the CREB homodimer (Craig et al., 2001; Schumacher et al., 2000). As shown above, lithium and magnesium had opposing effects on the interaction between CREB and TORC1 in GST pull-down assays. To investigate the putative role of the lysine residue at position 290 of CREB for the lithium effect on the cAMP-induced CREB-directed transcriptional activity, the CREB mutants CREB-K290E and CREB-K290A were generated by site-directed mutagenesis. The positively charged lysine (K) residue at position 290 was substituted with the negatively charged glutamate (E) residue in the CREB mutant CREB-K290E, whereas CREB-K290A contains the neutral alanine (A) residue at position 290 (Figure 24).

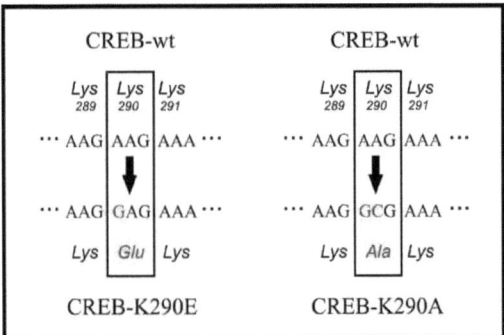

Figure 24: Mutations of the lysine residue at position 290 of CREB.
The lysine residue at position 290 of CREB was mutated by site-directed mutagenesis using primerless PCR. To substitute lysine (K) with glutamate (E) the codon AAG was changed to GAG as shown on the left. The substitution with alanine (A) was done by a codon change from AAG to GCG as demonstrated on the right.

13. DNA-binding of CREB-K290E

To explore the ability of the CREB mutant CREB-K290E to bind to the CRE as specific DNA binding site, electrophoretic mobility shift assays (EMSAs) were performed. GST-CREB-wt, GST-CREB-K290E, and GST alone were expressed in E.coli and purified by affinity chromatography using glutathione agarose. The proteins were eluted from the agarose and the concentration was determined photometrically. Figure 25A demonstrates equal quality of the eluted proteins by SDS-PAGE. GST-fusion proteins were employed together with the [^{32}P]-labeled double-stranded oligomer containing the rat *somatostatin* promoter CRE. Figure 25B shows a typical gel. Compared to the free [^{32}P]somCRE, the incubation with GST-CREB-wt led to a clear band shift (Figure 25B). In contrast, binding of GST-CREB-K290E was markedly reduced compared to GST-CREB-wt (Figure 25B). This indicates a weaker interaction between consensus CRE element and the mutant.

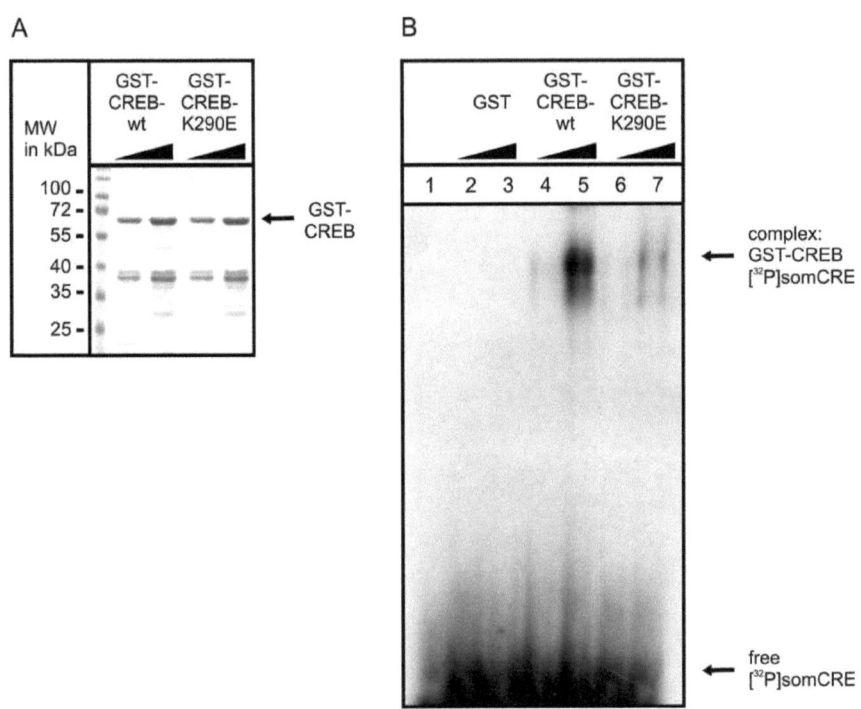

Figure 25: Effect of the mutation K290E on the DNA-binding ability of CREB.
GST-fusion proteins GST-CREB wild-type (wt) and mutant GST-CREB-K290E, and, for control, GST alone were expressed and purified from E.coli.
Figure 25A: Coomassie-stained SDS-PAGE of GST-fusion proteins. Proteins (each 3 μg and 6 μg) were electrophoresed by SDS-PAGE followed by Coomassie staining of the gel. GST-CREB-wt and GST-CREB-K290E showed an apparent molecular weight of 70 kDa and were of equal quality.
Figure 25B: Reduced DNA-binding of CREB-K290E in EMSAs. A typical gel is shown. 1.5 μg or 3 μg of GST, GST-CREB-wt, or GST-CREB-K290E, respectively, were used per reaction with [^{32}P]-labeled somCRE, containing the CRE of the rat *somatostatin*-gene promoter (somCRE). The free probe and the complex GST-CREB:[^{32}P]somCRE are indicated.
Lane 1: free [^{32}P]somCRE;
Lane 2: 1.5 μg GST and [^{32}P]somCRE;
Lane 3: 3 μg GST and [^{32}P]somCRE;
Lane 4: 1.5 μg GST-CREB-wt and [^{32}P]somCRE;
Lane 5: 3 μg GST-CREB-wt and [^{32}P]somCRE;
Lane 6: 1.5 μg GST-CREB-K290E and [^{32}P]somCRE;
Lane 7: 3 μg GST-CREB-K290E and [^{32}P]somCRE.
For the control GST no band shift of [^{32}P]somCRE was detectable. GST-CREB-wt showed a distinct band shift of [^{32}P]somCRE depending on the amount of protein used. GST-CREB-K290E exhibited markedly reduced amounts of shifted probe compared to GST-CREB-wt.

14. Effect of K290E and K290A mutations on CREB transcriptional activities

14.a Effect of the K290E mutation on CREB transcriptional activity under basal conditions and after stimulation by KCl and forskolin

To investigate the transcriptional activity of CREB-K290E, the GAL4 system was employed. The fusion of the transcription factor to the DNA binding domain of GAL4 allows the investigation independent of the DNA binding abilities of the protein. HIT-T15 cells were transiently transfected with the luciferase-reporter gene G5E1B-Luc coding for luciferase under control of five repeats of the GAL4-binding site. Cotransfected was an expression plasmid encoding either CREB-wt or CREB-K290E fused to the DNA-binding domain of GAL4 (Figure 26, top). The cells were treated with 45 mM KCl, 10 µM FSK, or the combination of both. As shown in Figure 26A, GAL4-CREB-K290E exhibited a markedly increased basal activity of 480.38 ± 59.47% compared to GAL4-CREB-wt ($p<0.001$). To indicate the specific effects of the respective treatment on the transcriptional activity, basal activity was set as 100% for each protein (Figure 26B). The treatment with 45 mM KCl increased the transcriptional activity of GAL4-CREB-wt 5-fold to 510.39 ± 107.44% ($p<0.004$), and treatment with 10 µM FSK enhanced the transcriptional activity 22-fold to 2206.66 ± 503.72% ($p<0.002$). The combination of KCl and FSK synergistically increased the activity of GAL4-CREB-wt 150-fold to 15058.71 ± 2474.20% ($p<0.001$; Figure 26B). Compared to GAL4-CREB-wt, the mutant GAL4-CREB-K290E exhibited virtually similar transcriptional activation by treatment with KCl, FSK, and the combination of KCl and FSK (Figure 26B). Statistical analysis of the data by two-way ANOVA revealed specific effects of the treatment with $p<0.0001$, but GAL4-CREB-wt and GAL4-CREB-K290E did not differ in the response to the treatment.

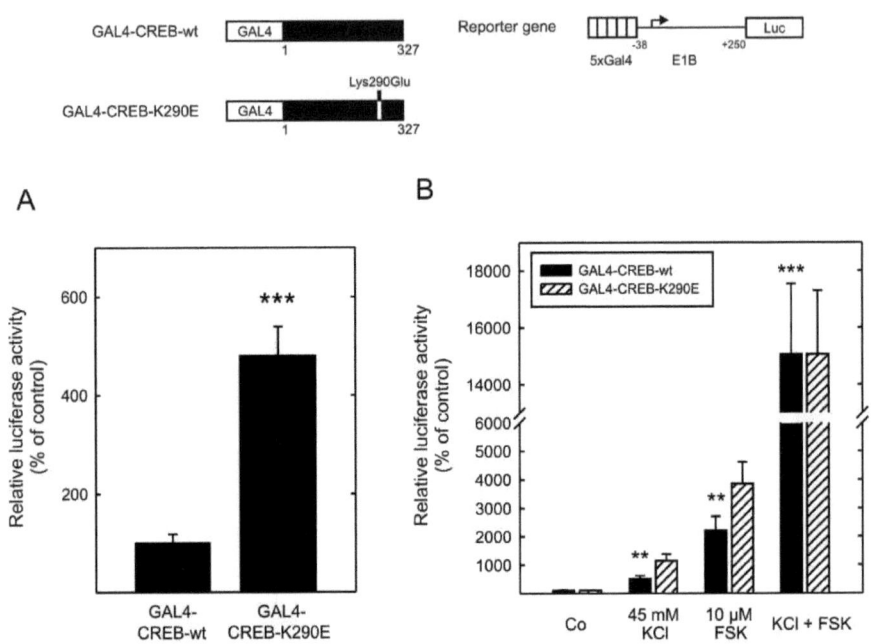

Figure 26: Effect of the K290E mutation on CREB transcriptional activity under basal conditions and after stimulation by KCl and forskolin.

HIT-T15 cells were transiently transfected with the luciferase-reporter gene G5E1B under control of five repeats of the GAL4-binding site. Cotransfected was an expression construct encoding CREB wild-type (wt) or mutant CREB-K290E fused to the DNA-binding domain of GAL4. A schematic illustration of the constructs is shown above the graphs. The cells were treated with 45 mM KCl for 6 h, with 10 µM forskolin (FSK) for 6 h, or with the combination of both, and luciferase activity was determined. Relative luciferase activity values are means ± SEM of three independent experiments performed in duplicate, and are expressed in percent of control without treatment.
Figure 26A: Comparison of the basal activity of GAL4-CREB wt and GAL4-CREB-K290E. GAL4-CREB-K290E exhibited 5-fold increased basal activity compared to GAL4-CREB-wt.
Figure 26B: Effects of treatment on the transcriptional activity of GAL4-CREB-wt and GAL4-CREB-K290E. To compare the effects of KCl and FSK the basal activity set as 100% for both GAL4-CREB-wt and GAL4-CREB-K290E. The treatment with 45 mM KCl or 10 µM FSK alone increased the basal activity of GAL4-CREB-wt. The combination of KCl and FSK enhanced the activity synergistically. The effects of the treatments on the activity of GAL4-CRE-K290E were comparable to wild-type.
Statistical analysis was performed by two-way ANOVA followed by Student's t-test: ***$p<0.001$, **$p<0.025$.

14.b Effect of K290E and K290A mutations on basal CREB transcriptional activity and on its stimulation by lithium in the presence of cAMP

HIT-T15 cells were transiently transfected with the luciferase-reporter gene G5E1B-Luc under control of five repeats of the GAL4 binding site. Cotransfected was an expression construct coding for CREB-wt, CREB-K290E, or CREB-K290A fused to the DNA-binding domain of GAL4 (Figure 27). The cells were treated with 20 mM LiCl, with 1 mM 8-bromo-cAMP, or with the combination of both and luciferase activity was determined. Compared to GAL4-CREB-wt, the basal activity of GAL-CREB-K290E and GAL4-CREB-K290A was increased 4-fold to 425.60 ± 32.23% and 397.99 ± 22.27% ($p<0.001$ and $p<0.0001$), respectively (Figure 27A). Statistical analysis by two-way ANOVA demonstrated a general difference between the proteins with $p<0.0001$. To compare treatment effects, the basal activity was set as 100% for each protein (Figure 27B). The treatment with 20 mM LiCl did not influence the transcriptional activity of GAL4-CREB-wt, GAL4-CREB-K290E, or GAL4-CREB-K290A. The treatment with 1 mM 8-bromo-cAMP increased the activity of GAL4-CREB-wt 3.2-fold to 324.06 ± 37.49% ($p<0.002$), whereas the activity of GAL4-CREB-K290E and GAL4-CREB-K290A was increased 9.4-fold to 941.85 ± 118.63% and 6.8-fold to 681.20 ± 147.50%, respectively (Figure 27B). The differences in responsiveness of the GAL4-CREB variants to treatment with 8-bromo-cAMP were confirmed by two-way ANOVA with $p<0.004$, but GAL4-CREB-K290E did not differ from GAL4-CREB-K290A. The cAMP-induced transcriptional activity was further enhanced by lithium treatment. The increase was 3.4-fold ($p<0.013$), 2.65-fold ($p<0.006$), and 2.21-fold ($p<0.04$) for GAL4-CREB-wt, GAL4-CREB-K290E, and GAL4-CREB-K290A, respectively (Figure 27B). The statistical analysis by two-way ANOVA revealed no significant difference between the proteins for the effect of lithium on cAMP-induced transcriptional activity (compare Figure 27C).

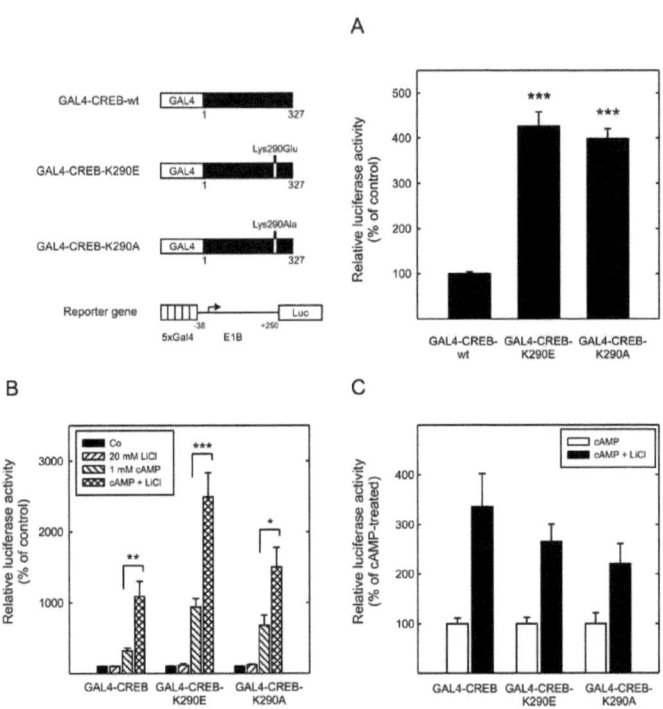

Figure 27: Effect of K290E and K290A mutations on basal CREB transcriptional activity and on its stimulation by lithium in the presence of cAMP in luciferase reporter-gene assays.

HIT-T15 cells were transiently transfected with the luciferase-reporter gene G5E1B under control of five repeats of the GAL4-binding site. Cotransfected was an expression plasmid encoding CREB wild-type (wt), CREB-K290E, or CREB-K290A fused to the DNA-binding domain of GAL4. A schematic illustration of the constructs is shown above the graphs. The cells were treated with 20 mM LiCl for 7 h, with 1 mM 8-bromo-cAMP for 6 h, or with the combination of both and luciferase activity was determined. Relative luciferase activity values are means ± SEM of two independent experiments performed in duplicate, and are expressed in percent of control without treatment. Statistical analysis was performed by two-way ANOVA followed by Student's t-test: ***$p<0.001$, **$p<0.025$; *$p<0.05$.

Figure 27A: Comparison of the basal activity of GAL4-CREB wt, GAL4-CREB-K290E and GAL4-CREB-K290A. GAL4-CREB-K290E and GAL4-CREB-K290A exhibited 4-fold increased basal activity compared to GAL4-CREB-wt.

Figure 27B: Effects of treatment on the transcriptional activity of GAL4-CREB-wt, GAL4-CREB-K290E and GAL4-CREB-K290A. To compare the effects the basal activity was set as 100% in each group. The treatment with 20 mM LiCl alone did not affect the activity of either CREB-wt or mutants. 8-bromo-cAMP increased the basal activity of GAL4-CREB-wt, which was stronger for the mutants. LiCl enhanced the transcriptional activity induced by 8-bromo-cAMP of CREB-wt and mutants similarly.

Figure 27C Induction by LiCl of cAMP-induced transcriptional activity of GAL4-CREB-wt, GAL4-CREB-K290E, and GAL4-CREB-K290A. The effects of LiCl treatment on the activity induced by 8-bromo-cAMP were comparable to the wild-type.

14.c Comparison of the expression level of GAL4-CREB wild-type and K290E/K290A mutants

The expression level of GAL4-CREB-wt, GAL4-CREB-R300A, GAL4-CREB-K290E, and GAL4-CREB-K290A was examined by Western blot. HIT-T15 cells were transfected with an expression construct encoding CREB-wt, CREB-R300A, CREB-K290E, or CREB-K290A fused to the DNA-binding domain of GAL4. Cell lysates were electrophoresed by SDS-PAGE. The proteins were transferred to a membrane and immunostained with the CREB-KID antibody. Figure 28 demonstrates a typical blot. GAL4-CREB proteins with an apparent molecular weight of 60 kDa, and endogenous CREB with an apparent molecular weight of 43 kDa are indicated respectively. The analysis revealed a similar expression of GAL4-CREB-wt and GAL4-CREB-R300A. Notably, the mutants GAL4-CREB-K290E and GAL4-CREB-K290A showed a reduced expression level compared to GAL4-CREB-wt (Figure 28).

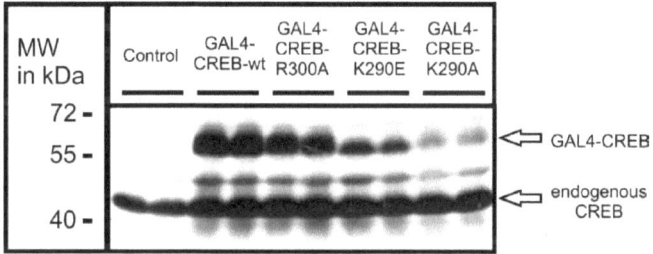

Figure 28: Comparison by Western blot of the expression levels of GAL4-CREB-wt and the mutants R300A, K290E, and K290A.
To compare the expression levels of GAL4-fusion proteins, HIT-T15 cells were transiently transfected with expression constructs encoding CREB wild-type (wt), CREB-R300A, CREB-K290E, or CREB-K290A fused to the DNA-binding domain of GAL4. Untransfected cells were used as control. Whole cell lysates were prepared from the cells by hot lysis and 15 µl per sample were elctrophoresed by SDS-PAGE. The proteins were blotted to a nitrocellulose membrane by semi-dry transfer and the membrane was probed with the CREB-KID antibody diluted 1:25000. A typical blot is shown. Endogenous CREB and the GAL4-CREB showed an apparent molecular weight of 43 kDa and 60 kDa, as indicated, respectively. GAL4-CREB-wt and GAL4-CREB-R300A showed similar expression levels, whereas lower expression of the mutants GAL4-CREB-K290E and GA4-CREB-K290A was evident.

14.d Effect of K290E and K290A mutations on the transcriptional activity conferred by TORC1 to the CREB bZip

14.d.I Effect of K290E and K290A mutations on the transcriptional activity conferred by TORC1 to the CREB bZip and its stimulation by lithium in the presence of cAMP

TORC1 was shown to confer the enhancement by lithium of the cAMP-induced CREB transcriptional activity (compare Figure 13). Now, the effect of the CREB-K290E and CREB-K290A mutations was investigated. HIT-T15 cells were transiently transfected with the luciferase-reporter gene G5E1B-Luc under control of five repeats of the GAL4-binding site and an expression plasmid encoding the full-length human TORC1. Cotransfected was an expression construct encoding CREB-bZip-wt, CREB-bZip-K290E, or CREB-bZip-K290A fused to the DNA-binding domain of the yeast transcription factor GAL4 (Figure 29, top). The cells were treated with 20 mM LiCl, with 2 mM 8-bromo-cAMP, or with the combination of both. The GAL4-bZip-wt alone exhibited virtually no transcriptional activity as was demonstrated before (Figure 29A, compare Figure 13A). Upon overexpression of TORC1 the basal activity of GAL4-bZip-wt was markedly increased ($p<0.0001$). Compared to GAL4-bZip-wt, the mutants GAL4-bZip-K290E and GAL4-bZip-K290A exhibited 3-fold increased basal transcriptional activities with 304.83 ± 25.38% and 305.45 ± 20.55% ($p<0.001$ and $p<0.0001$), respectively, when TORC1 was overexpressed (Figure 29A). The difference in basal activity was confirmed with $p<0.0001$ by two-way ANOVA. The treatment with 20 mM LiCl slightly increased the transcriptional activity of GAL4-bZip-wt as well as the mutants GAL4-bZip-K290E and GAL4-bZip-K290A. The treatment with 2 mM 8-bromo-cAMP enhanced the transcriptional activity of GAL4-bZip-wt after overexpression TORC1 2.5-fold to 252.03 ± 19.18% ($p<0.001$), whereas the transcriptional activity of GAL4-bZip-K290E and GAL4-bZip-K290A was increased 3.37-fold to 1011.92 ± 36.61% ($p<0.0001$) and 2.76-fold to 843.80 ± 54.16% ($p<0.0001$), respectively (Figure 29A). The proteins did not differ with respect to the transcriptional activity induced by 8-bromo-cAMP as revealed by two-way ANOVA. As shown before for the GAL4-bZip-wt, the treatment with 20 mM LiCl further increased the transcriptional activity induced by 8-bromo-cAMP 2.32-fold ($p<0.005$) when TORC1 was overexpressed (Figure 29A, compare Figure 13A). Also for GAL4-bZip-K290E and for GAL4-bZip-K290A, lithium increased the cAMP-induced transcriptional activity 2.09-fold ($p<0.0001$) and 2.18-fold ($p<0.002$), respectively (Figure 29A and 29B). Regarding the effect of lithium on cAMP-induced transcriptional activity mediated by TORC1, the GAL4-bZip-wt and mutants did not differ from each other, as confirmed by two-way ANOVA (Figure 29B).

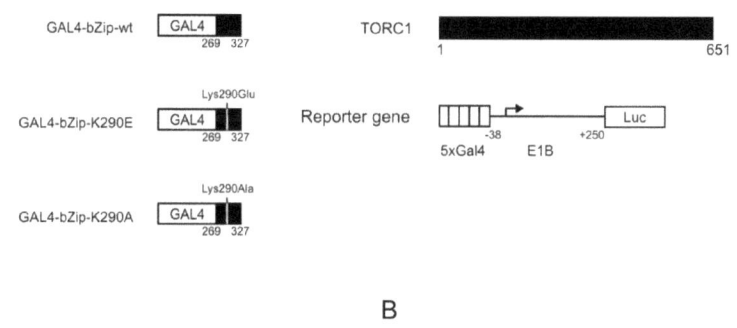

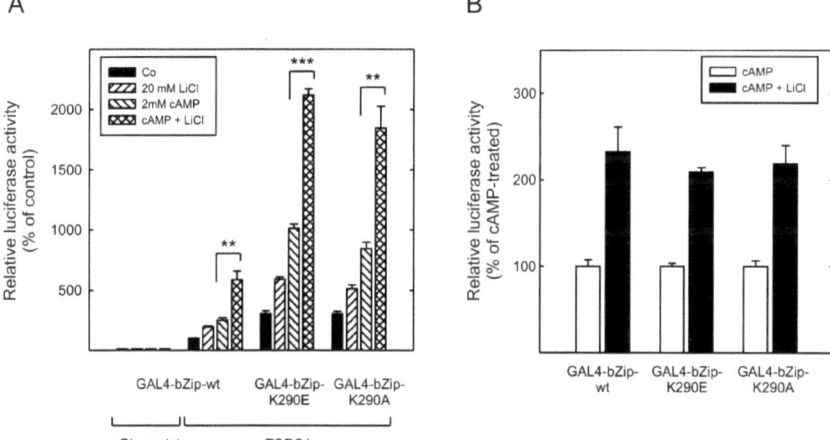

Figure 29: Effects of lithium and cAMP on the transcriptional activity of the CREB bZip conferred by TORC1 in luciferase reporter-gene assays – comparison between CREB bZip wild-type and the mutants K290E and K290A.

HIT-T15 cells were transiently transfected with the luciferase-reporter gene G5E1B under control of five repeats of the GAL4-binding site and with an an expression plasmid encoding full-length human TORC1. Cotransfected was an expression plasmid encoding the basic leucine zipper (bZip) wild-type (wt), bZip-K290E, or bZip-K290A fused to the DNA-binding domain of GAL4. A schematic illustration of the constructs is shown above the graphs. The cells were treated with 20 mM LiCl for 7 h, with 2 mM 8-bromo-cAMP for 6 h, or with the combination of both and luciferase activity was determined. Relative luciferase activity values are means ± SEM of two independent experiments performed in duplicate, and are expressed in percent of GAL4-bZip-wt upon overexpression of TORC1. Statistical analysis was performed by two-way ANOVA followed by Student's t-test: ***p<0.001, **p<0.025.

Figure 29A: Comparison of the effects of treatment on the transcriptional activity of GAL4-bZip-wt, GAL4-bZip-K290E and GAL4-bZip-K290A upon overexpression of TORC1. The GAL4-bZip alone did not exhibit transcriptional activation (cotransfection of the empty vector pBluescript). Upon overexpression of TORC1, the basal activity was increased, which was 3-fold higher for the mutants compared to the GAL4-bZip-wt. The treatment with 20 mM LiCl alone or 2 mM 8-bromo-cAMP alone increased the activity of either GAL4-bZip-wt or mutants. The enhancement by LiCl of the transcriptional activity induced by 8-bromo-cAMP was evident for GAL4-bZip-wt and mutants.
Figure 29B: Comparison of the effects of LiCl on cAMP-induced transcriptional activity of GAL4-bZip-wt, GAL4-bZip-K290E, and GAL4-bZip-K290A. The enhancement by LiCl of the transcriptional activity induced by 8-bromo-cAMP was similar for the mutants compared to the wild-type.

14.d.II Specificity of the effect of TORC1 overexpression on GAL4-bZip transcriptional activity

To verify the specificity of the effect of TORC1 on the transcriptional activity of GAL4-bZip-wt it was overexpressed also with the GAL4 DNA-binding domain alone (GAL4-DBD). HIT-T15 cells were transiently transfected with the luciferase-reporter gene G5E1B encoding luciferase under control of five repeats of the GAL4-binding site and an expression plasmid for full-length human TORC1. Cotransfected was the GAL4-bZip or the GAL4-DBD (Figure 30, left). The cells were treated with 20 mM LiCl, with 1 mM 8-bromo-cAMP, or with the combination of both. As described above, the transcriptional activity of GAL4-bZip-wt was enhanced by overexpression of TORC1 and further increased upon the treatment with 8-bromo-cAMP and lithium (p<0.001; Figure 30, compare Figure 29A, and Figure 13A). In contrast, the GAL4-DBD alone exhibited virtually no transcriptional activity upon overexpression of TORC1 (Figure 30) as confirmed by two-way ANOVA with p<0.0001. This result argues against unspecific effects due to the GAL4 DNA-binding domain.

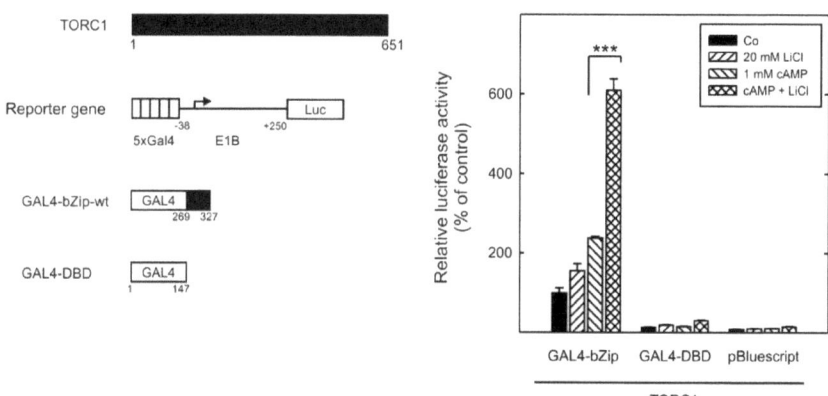

Figure 30: TORC1 overexpression specifically activates transcriptional activity of GAL4-bZip.

HIT-T15 cells were transiently transfected with the luciferase-reporter gene G5E1B under control of five repeats of the GAL4-binding site and with an expression plasmid encoding full-length human TORC1. Cotransfected was an expression plasmid encoding the CREB bZip wild-type (wt) fused to the DNA-binding domain of GAL4, or the GAL4 DNA-binding domain alone (GAL4-DBD). For control the empty vector pBluescript was cotransfected. A schematic illustration of the constructs is shown. The cells were treated with 20 mM LiCl for 7 h, with 1 mM 8-bromo-cAMP for 6 h, or with the combination of LiCl and 8-bromo-cAMP. Luciferase activity was determined. Relative luciferase activity values are means ± SEM of two independent experiments performed in duplicate, and are expressed in percent of GAL4-bZip-wt without treatment. Overexpression of TORC1 mediates transcriptional activation of the GAL4-bZip-wt, which was further increased by treatment with 8-bromo-cAMP and lithium. The GAL4-DBD alone did not exhibit transcriptional activation upon overexpression of TORC1 and was comparable to the control pBluescript. Statistical analysis was performed by two-way ANOVA followed by Student's t-test: ***$p<0.001$.

15. Effect of K290E and K290A mutations on the interaction of CREB with TORC1 and its stimulation by lithium

15.a Mammalian two-hybrid assay

At first the expression level of the VP16-fusion proteins and their nuclear localization was investigated by Western blot analysis of cytosolic and nuclear fraction of transiently transfected HIT cells. The samples were electrophoresed by SDS-PAGE; the proteins were transferred to a membrane and immunostained with an antibody recognizing VP16. Upon transfection of equal amount of DNA, the Western Blot analysis revealed the dissimilar expression level of the VP16-fusion proteins in nuclear fractions of HIT-T15 cells (Figure 31A). The mutants VP16-bZip-K290E and VP16-bZip-K290A were expressed at higher levels compared to VP16-bZip-wt, whereas the mutant VP16-bZip-R300A was not detectable (Figure 31A). The amount of DNA to be transfected was adjusted respectively to reach a similar expression level of the proteins which was controlled again by Western blot. Figure 31B shows a Western blot of cytosolic and nuclear fractions of HIT-T15 cells transfected with adjusted amounts of DNA. The expression of the VP16-fusion proteins reached similar levels in nuclear fractions (Figure 31B, upper panel).

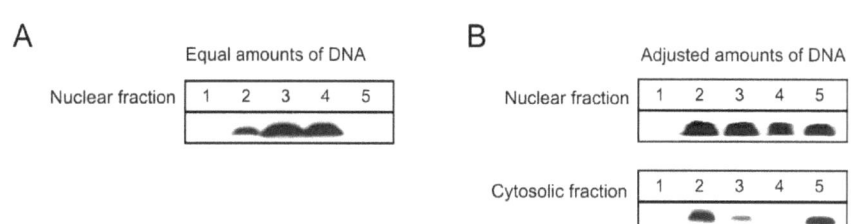

Figure 31: Comparison by Western blot of the expression levels of VP16-bZip wild-type and mutants in HIT-T15 cells.

HIT-T15 cells were transiently transfected with expression plasmids encoding bZip wild-type (wt), bZip-K290E, bZip-K290A, or bZip-R300A fused to the viral protein VP16. Cytosolic and nuclear proteins were extracted from the cells. 200 µg or 150 µg of nuclear or cytosolic proteins, respectively, were electrophoresed by SDS-PAGE. The proteins were transferred to a nitrocellulose membrane by semi-dry transfer and the membrane was probed with the VP16(1-21) antibody diluted 1:200.
Lane 1: untransfected cells
Lane 2: VP16-bZip-wt
Lane 3: VP16-bZip-K290E
Lane 4: VP16-bZip-K290A
Lane 5: VP16-bZip-R300A
Figure 31A Expression levels of VP16-fusion proteins in nuclear fractions of HIT-T15 cells upon transfection of equal amounts of DNA per group. The expression of the mutants VP16-bZip-K290E and VP16-bZip-K290A was markedly increased compared to VP16-bzip-wt, and VP16-bZip-R300A was not detectable.
Figure 31B Expression levels of VP16-fusion proteins in nuclear and cytosolic fractions of HIT-T15 cells upon transfection of different amounts of DNA per group. Adjustment of DNA amounts used for transfection of HIT-T15 cells led to similar expression of VP16-fusion proteins. Compared to the cytosolic protein fractions, VP16-fusion proteins are mainly localized in the nucleus.

For the mammalian two-hybrid assay, HIT-T15 cells were transiently transfected with the luciferase-reporter gene G5E1B-Luc controlled by five repeats of the GAL4-binding site and an expression construct for the first 44 amino acids of TORC1 fused to the DNA binding domain of GAL4. Cotransfected was an expression construct encoding the bZip-wt, bZip-K290E, bZip-K290A, or bZip-R300A fused to the viral protein VP16 (Figure 32, left). The amount of DNA was adjusted to assure equal expression of the VP16-fusion proteins. For control conditions VP16 alone was used. The cells were treated with 20 mM LiCl. Compared to the control VP16, the expression of VP16-bZip-wt increased the basal activity of GAL4-TORC1$_{1-44}$ 1.5-fold to 148.16 ± 10.91% ($p<0.002$). The expression of the mutants VP16-bZip-K290E and VP16-bZip-K290A led to 1.3-fold (132.90 ± 11.56%, $p<0.023$) and 1.25-fold (125.30 ± 7.66%, $p<0.015$) increased basal activity of GAL4-TORC1$_{1-44}$, respectively (Figure 32). The mutant VP16-bZip-R300A did not show significant difference from the control VP16 alone. Treatment with 20 mM LiCl increased the luciferase activity 4.7-fold to 705.47 ± 72.04% ($p<0.0001$) for the VP16-bZip-wt (Figure 32). Lithium treatment enhanced the luciferase activity for VP16-bZip-K290E 1.66-fold to 222.38 ± 21.20% ($p<0.005$) and for VP16-bZip-K290A 2.18-fold to 278.20 ± 34.09% ($p<0.002$) as shown in Figure 32. The statistical analysis by two-way ANOVA revealed a significant effect of LiCl treatment upon expression of VP16-bZip-K290E and VP16-bZip-K290A with $p<0.023$ compared to VP16 alone, which was not present for the mutant VP16-bZip-R300A. Noteworthy, the effect of LiCl treatment was significantly lower for the mutants VP16-bZip-K290E and VP16-bZip-K290A compared to VP16-bZip-wt (two-way ANOVA: $p<0.0001$). The proteins VP16-bZip-K290E and VP16-bZip-K290A did not differ from each other with respect to the effect of LiCl treatment.

Results

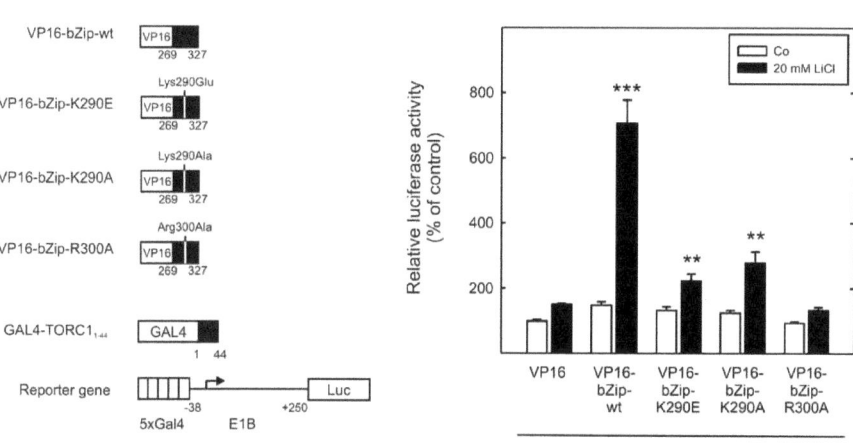

Figure 32: Effect of K290E and K290A mutations on the interaction of TORC1 with CREB and its stimulation by lithium as revealed in a mammalian two-hybrid assay.

HIT-T15 cells were transiently transfected with the luciferase-reporter gene G5E1B under control of five repeats of the GAL4-binding site and an expression construct encoding the first 44 amino acids of TORC1 fused to the DNA-binding domain of GAL4. Cotransfected was an expression plasmid encoding the CREB basic leucine zipper (bZip) wild-type (wt), bZip-K290E, bZip-K290A, or bZip-R300A fused to the viral protein VP16, or VP16 alone. A schematic illustration of the constructs is shown. The cells were treated with 20 mM LiCl for 30 h and luciferase activity was determined. Relative luciferase activity values are means ± SEM of two independent experiments performed in triplicate. Relative luciferase units (RLU) are expressed in percent of the control VP16. Compared to VP16 alone, the basal activity of GAL4-TORC1$_{1-44}$ was increased upon expression of VP16-bZip-wt. 20 mM LiCl strongly increased the luciferase activity. Expression of VP16-bZip-K290E or VP16-bZip-K290A also increased the basal activity of GAL4-TORC1$_{1-44}$ comparable to the wild-type. The enhancement by lithium was markedly reduced but significantly different from the control VP16 alone. VP16-bZip-R300A was not different from the control VP16 alone. Statistical analysis was performed by two-way ANOVA followed by Student's t-test: ***$p<0.001$, **$p<0.025$.

15.b *In vitro* GST pull-down assay

15.b.I Effect of lithium

To investigate the effect of lithium on the interaction between the full-length CREB-K290E or CREB-K290A and TORC1 under cell-free conditions, GST pull-down assays were performed. Recombinant GST-CREB-wt, GST-CREB-K290E, and GST-CREB-K290A (Figure 33A) and [^{35}S]TORC1 (Figure 8) were employed with and without LiCl in a binding reaction. The amount of [^{35}S]-labeled TORC recovered from GST-CREB-wt was determined by SDS-PAGE and subsequent autoradiography. A typical gel is shown below the graph (Figure 33B). Compared to GST-CREB-wt the control GST alone exhibited no remarkable binding of [^{35}S]TORC1 (8.51 ± 6.66% of GST-CREB-wt). The amount of [^{35}S]TORC1 recovered from GST-CREB-K290E was markedly reduced with 27.92 ± 4.14% of GST-CREB-wt ($p<0.001$; Figure 33B). The mutant GST-CREB-K290A also showed reduced binding of [^{35}S]TORC1 with 56.79 ± 5.12% of GST-CREB-wt ($p<0.004$). The presence of 20 mM LiCl in the reaction buffer increased the amount of [^{35}S]TORC1 recovered from GST-CREB-wt to 139.20 ± 6.66% ($p<0.01$) compared to the untreated control (Figure 33B). The recovery of [^{35}S]TORC1 from the mutant GST-CREB-K290E was increased 1.77-fold to 49.33 ± 10.13% of GST-CREB-wt by 20 mM LiCl ($p<0.04$). Also for GST-CREB-K290A an enhanced binding of [^{35}S]TORC1 was observed at 20 mM LiCl (Figure 33B). The amount of [^{35}S]TORC1 was increased 1.43-fold to 81.16 ± 6.39% of GST-CREB-wt. Statistical analysis by two-way ANOVA confirmed significant effects by LiCl on the binding of [^{35}S]TORC1 with $p<0.0001$, but GST-CREB-wt and GST-CREB-K290E or GST-CREB-K290A did not differ from each other in this aspect.

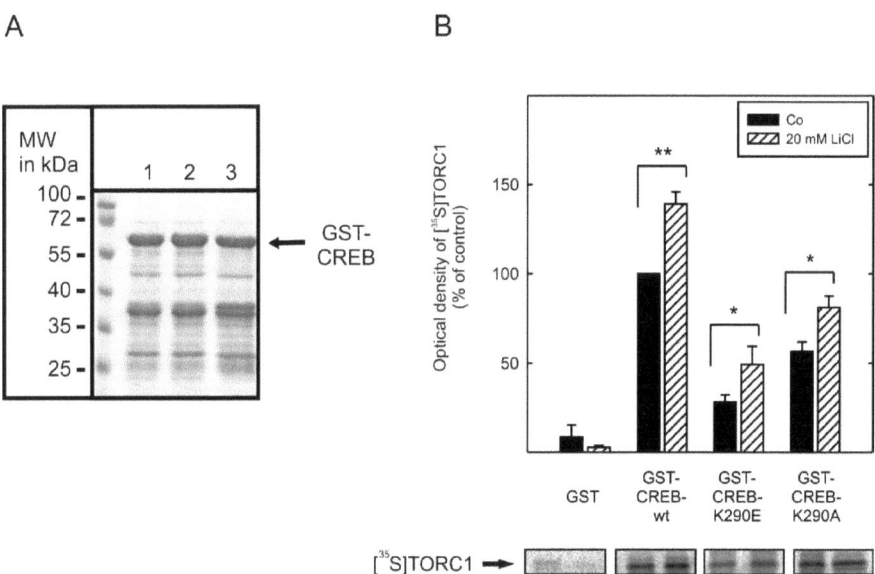

Figure 33: Effect of K290E and K290A mutations on the interaction of TORC1 with CREB and its stimulation by lithium as revealed in the GST pull-down assay.

GST-CREB wild-type (wt), GST-CREB-K290E and GST-CREB-K290A, and, for control, GST alone were purified from *E.coli*. TORC1 was labeled with [^{35}S]methionine by *in vitro* transcription and translation. The amount of [^{35}S]TORC1 recovered from GST-fusion proteins was analyzed by SDS-PAGE followed by autoradiography and densitometric analysis of the bands corresponding to [^{35}S]TORC1. A typical gel is shown below the graph.
Figure 33A: Coomassie-stained SDS-PAGE gel of GST-CREB-wt, GST-CREB-K290E, and GST-CREB-K290A. The proteins were expressed in *E.coli*, purified by affinity chromatography and immobilized on glutathione agarose. The samples were electrophoresed by SDS-PAGE and the gel was stained with Coomassie. Equal quality and amounts of proteins to be used in GST pull-down assays were evident.
Figure 33B: Lithium increased the interaction between [^{35}S]TORC1 and GST-CREB independent of the mutation of K290 of CREB. The data are mean values ± SEM of three independent experiments and are expressed in percent of GST-CREB-wt. GST alone exhibited no remarkable binding of [^{35}S]TORC1 compared to GST-CREB-wt. 20 mM LiCl increased the amount of [^{35}S]TORC1 recovered from GST-CREB-wt. The mutant GST-CREB-K290E exhibited strongly reduced binding of [^{35}S]TORC1. In the presence of 20 mM LiCl the amount of [^{35}S]TORC1 recovered from GST-CREB-K290E was increased. The mutant GST-CREB-K290A also showed reduced binding of [^{35}S]TORC1. 20 mM LiCl increased the binding of [^{35}S]TORC1 to GST-CREB-K290A. The effect of LiCl on the binding of [^{35}S]TORC1 was equivalent for GST-CREB-wt and the mutants. Statistical analysis was performed by two-way ANOVA followed by two-sided paired Student's *t*-test: **$p<0.025$; *$p<0.05$.

15.b.II Effect of magnesium

It was shown before that magnesium inhibited the interaction between CREB and TORC1 in GST pull-down assays (Figure 23). In this context, the effect of the K290E mutation on the inhibition by magnesium of the interaction between CREB and TORC1 was investigated. Recombinant GST-CREB-wt and GST-CREB-K290E (Figure 33A) and [^{35}S]TORC1 (Figure 8) were employed with and without $MgCl_2$ in a binding reaction. The amount of [^{35}S]-labeled TORC recovered from GST-CREB was determined by SDS-PAGE and subsequent autoradiography. To compare effects of magnesium ions on the binding of [^{35}S]TORC1 the group without magnesium was set as 100% for both GST-CREB-wt and GST-CREB-K290E. At 5 mM $MgCl_2$ the amount of [^{35}S]TORC1 recovered from GST-CREB-wt was reduced to 60.63 ± 9.92% of the untreated control (p<0.017, Figure 34). Comparable to the wild-type, the binding of [^{35}S]TORC1 to GST-CREB-K290E was reduced to 70.75 ± 18.41% of the untreated control. At 20 mM $MgCl_2$ the binding of [^{35}S]TORC1 was reduced to 35.63 ± 9.46% (p<0.003) for GST-CREB-wt and to 33.66 ± 13.92% for GST-CREB-K290E (p<0.009; Figure 34). The analysis by two-way ANOVA confirmed significant effects of $MgCl_2$ with p<0.001, but GST-CREB-K290E did not differ from GST-CREB-wt in this aspect.

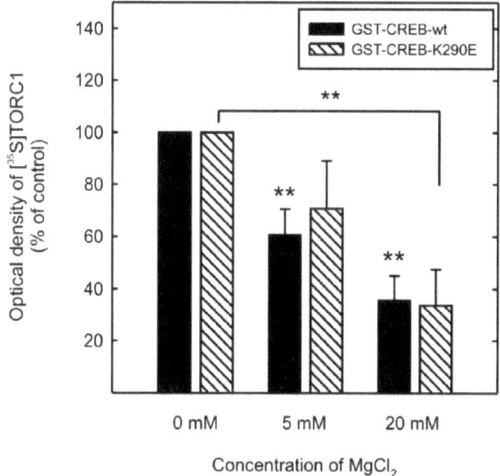

Figure 34: Effect of K290E mutation on the inhibition by magnesium of the interaction between TORC1 and CREB in the GST pull-down assay.

GST-CREB wild-type (wt) and GST-CREB-K290E were purified from *E.coli*. TORC1 was labeled with [^{35}S]methionine by *in vitro* transcription and translation. The amount of [^{35}S]TORC1 recovered from GST-fusion proteins was analyzed by SDS-PAGE followed by autoradiography and densitometric analysis of the bands. The data are mean values ± SEM of three independent experiments and are expressed in percent of control per group. MgCl$_2$ significantly reduced the amount of [^{35}S]TORC1 recovered from GST-CREB-K290E comparable with GST-CREB-wt. Statistical analysis was performed by two-way ANOVA followed by two-sided paired Student's *t*-test: **$p<0.025$.

16. Effect of K290E and K290A mutations on the recruitment of TORC1 to CREB bZip at the promoter

To examine effects of lithium on the recruitment of TORC1 to CREB bZip wild-type, bZip-K290E, and bZip-K290A at the G5E1B-Luc promoter, chromatin immunoprecipitation (ChIP) assays were performed. HIT-T15 cells were transiently transfected with an expression construct encoding FLAG-tagged TORC1 and with the expression construct G5E1B-Luc under control of five repeats of the GAL4-binding site. Cotransfected was an expression construct for the CREB-bZip-wt, CREB-bZip-K290E, or CREB-bZip-K290A fused to the DNA-binding domain of GAL4 (Figure 35). To control for the specificity of the assay the empty vector pBluescript was cotransfected. The cells were treated with 2 mM 8-bromo-cAMP or with the combination of 2 mM 8-bromo-cAMP and 20 mM LiCl. Subsequently, proteins and DNA were crosslinked and the protein-DNA complexes were precipitated by the FLAG M2 antibody. The DNA was extracted from the precipitates and the amount was quantified by real-time PCR specific for the G5E1B promoter. Compared to the GAL4-bZip-wt at G5E1B-Luc, the control pBluescript did virtually exhibit no specific signal in the real-time PCR (Figure 35A). Upon treatment with 8-bromo-cAMP, the promoter occupancy of FLAG-TORC1 with GAL4-bZip-K290E and GAL4-bZip-K290A was increased 2-fold (208.90 ± 26.92%, p<0.001) and 2.4-fold (236.90 ± 57.83%, p<0.03) compared to GAL4-bZip-wt (Figure 35B). This difference was confirmed by one-way ANOVA with p<0.032. As demonstrated in Figure 35C, the cotreatment with 2 mM 8-bromo-cAMP and 20 mM LiCl increased the amount of G5E1B-Luc DNA precipitated with FLAG-TORC1 (two-way ANOVA: p<0.0001) . For the control, GAL4-bZip-wt, LiCl treatment increased the promoter occupancy of FLAG-TORC1 2.4-fold (237.26 ± 63.12%, p<0.042), for GAL4-bZip-K290E 3-fold (314.54 ± 63.28%, p<0.005) and for GAL4-bZip-K290A 2-fold (200.60 ± 29.66, p<0.021). The analysis by two-way ANOVA revealed no difference between GAL4-bZip-wt, GAL4-bZip-K290E, and GAL4-bZip-K290A with respect to the effects elicited by lithium.

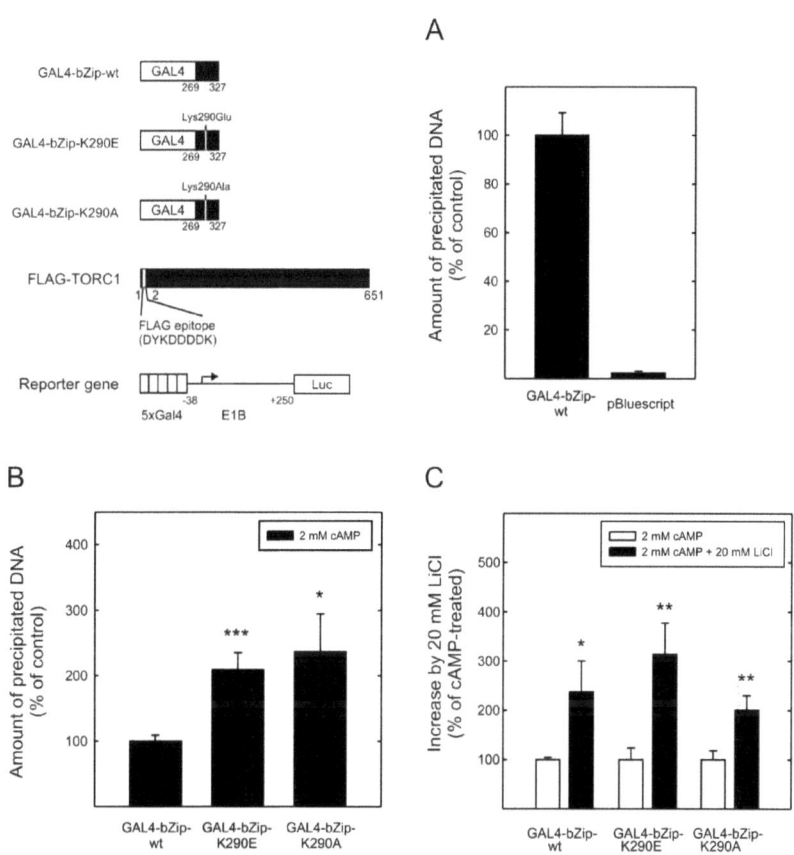

Figure 35: Effects of lithium on the recruitment of TORC1 to GAL4-bZip at the promoter – comparison between CREB bZip wild-type and the mutants K290E and K290A in chromatin immunoprecipitation assays.

For chromatin immunoprecipitation (ChIP) assays HIT-T15 cells were transiently transfected with an expression plasmid encoding the FLAG-tagged TORC1 and the reporter plasmid G5E1B controlled by a minimal promoter consisting of 5 repeats of the GAL4-binding site. Cotransfected was an expression construct encoding the bZip-wt, bZip-K290E, or bZip-K290A fused to the DNA-binding domain of GAL4. A schematic illustration of the constructs is shown. For control conditions the empty vector pBluescript was cotransfected. The cells were treated for 30 min with 2 mM 8-bromo-cAMP, or with the combination of 8-bromo-cAMP and 20 mM LiCl. DNA and proteins were crosslinked with 1% formaldehyde, the complexes were precipitated by use of the anti-FLAG M2 antibody, and the amount of precipitated DNA was determined by quantitative real-time PCR specific for the G5E1B promoter. Mean values ± SEM of 5 independent experiments are shown. Values are the amount of precipitated DNA molecules expressed in percent of control.
Figure 35A Specificity control of the assay. The control pBluescript exhibited no specific signal in quantitative real-time PCR.

Figure 35B Recruitment of FLAG-TORC1 to GAL4-bZip at the promoter. Compared to GAL4-bZip-wt the mutants K290E and K290A showed significantly increased recruitment of FLAG-TORC1 as revealed by increased amounts of precipitated DNA.
Figure 35C Lithium enhanced the recruitment of FLAG-TORC1 to the promoter. LiCl combined with 8-bromo-cAMP increased the amount of FLAG-TORC1 recruited to GAL4-bZip-wt. This effect was also observed for the mutants GAL4-bZip-K290E and GAL4-bZip-K290A. Statistical analysis was performed by two-way ANOVA followed by Student's t-test: ***$p<0.001$; **$p<0.025$; *$p<0.05$.

17. Stimulation by lithium of CREB/TORC-directed gene transcription induced by cAMP– evidence at native human promoters

17.a Effects of lithium at the human *fos*-gene promoter

To investigate the effects of lithium on cAMP-induced CREB-directed transcriptional activity at the *fos*-gene promoter, a luciferase reporter-gene construct (-711Fos-Luc) controlled by the promoter region -711 to +45 of the human *fos*-gene was employed for transient transfection of HIT-T15 cells. To examine the contribution of the CRE at position -60, the luciferase-reporter gene -711FosCREm-Luc was used in which the CRE was mutated by site-directed mutagenesis (Figure 36, top). Cotransfected was an expression plasmid encoding the full-length human TORC1 (Figure 36, top). The cells were treated with 20 mM LiCl, with 1 mM 8-bromo-cAMP, or with the combination of both. Under control conditions 20 mM LiCl, 1 mM 8-bromo-cAMP, or the combination of both increased the promoter activity of -711Fos-Luc only in tendency (Figure 36). The overexpression of TORC1 did not affect the basal promoter activity of -711Fos-Luc. Treatment with LiCl alone or 8-bromo-cAMP alone did not increase the activity. In contrast, 20 mM LiCl enhanced the promoter activity of -711Fos-Luc in the presence of 8-bromo-cAMP 1.7-fold to 458.89 ± 56.47% ($p<0.02$) when TORC1 was overexpressed (Figure 36). The mutation of the CRE at position -60 of -711FosCREm-Luc resulted in reduced basal promoter activity. Under control condition -711FosCREm-Luc exhibited 55.55 ± 10.63% of -711Fos-Luc basal activity (Figure 36). The treatment with LiCl and 8-bromo-cAMP exerted moderate effects on the promoter activity. Upon overexpression of TORC1, LiCl enhanced the promoter activity in the presence of 8-bromo-cAMP 1.77-fold to 256.98 ± 32.60% ($p<0.003$; Figure 36). The statistical analysis by two-way ANOVA confirmed significant effects of LiCl in dependence of the overexpression of TORC1 with $p<0.002$, but the promoter activities of -711Fos-Luc and -711FosCREm-Luc were not different from each other in this respect.

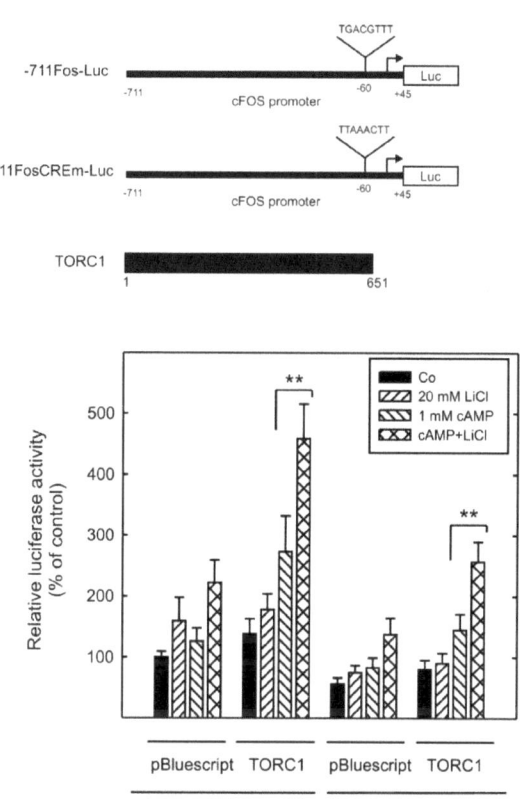

Figure 36: Effects of lithium and cAMP on human *fos*-gene transcription in luciferase reporter-gene assays.

HIT-T15 cells were transiently transfected with the luciferase-reporter gene -711Fos-Luc, comprising the promoter region -711 to +45 of the human *fos*-gene promoter, or -711FosCREm-Luc in which the CRE at position -60 was mutated. Cotransfected was an expression plasmid encoding full-length human TORC1. For control the empty expression vector pBluescript was cotransfected. A schematic illustration of the constructs is shown. The cells were treated with 20 mM LiCl for 7 h, with 1 mM 8-bromo-cAMP for 6 h, or with the combination of LiCl and 8-bromo-cAMP. Luciferase activity was determined. Relative luciferase activity values are means ± SEM of three independent experiments performed in duplicate, and are expressed in percent of basal activity of -711Fos-Luc. Statistical analysis was performed by two-way ANOVA followed by Student's *t*-test: **$p<0.025$.

Under control condition, treatment with LiCl and 8-bromo-cAMP alone or in combination did not significantly affect the promoter activity of -711Fos-Luc. Upon overexpression of TORC1, LiCl increased the promoter activity in the presence of 8-bromo-cAMP. Mutation of the CRE decreased the promoter activity of -711FosCREm-Luc under control conditions. Upon overexpression of TORC1 an increase of the 8-bromo-cAMP induced promoter activity of -711FosCREm-Luc was evident.

17.b Effects of lithium at the human *BDNF(exon IV)*-gene promoter

To investigate the effects of lithium on cAMP-induced CREB-directed transcriptional activity at the *BDNF(exon IV)*-gene promoter, a luciferase reporter-gene construct (BDNF4-Luc) controlled by the promoter region -242 to +306 of the exon IV of the human *BDNF*-gene was employed for transient transfection of HIT-T15 cells. To examine effects elicited by CREB, HIT-T15 cells were transfected with the luciferase-reporter gene BDNF4CREm-Luc in which the CRE was mutated (Figure 37, top). Cotransfected was an expression construct encoding the full-length human TORC1. The cells were treated with 20 mM LiCl, with 1 mM 8-bromo-cAMP, or with the combination of both. Under control conditions 20 mM LiCl, 1 mM 8-bromo-cAMP, or the combination of LiCl and 8-bromo-cAMP did not enhance the promoter activity (Figure 37). The overexpression of TORC1 increased the basal promoter activity 2-fold to 209.83 ± 12.65% ($p<0.0001$; Figure 37). Under these conditions the treatment with LiCl or 8-bromo-cAMP alone did not affect the activity. The treatment with LiCl increased the transcriptional activity in the presence of 8-bromo-cAMP 1.25-fold to 286.80 ± 20.50% ($p<0.011$) when TORC1 was overexpressed (Figure 37). Compared to BDNF4-Luc the basal promoter activity of BDNF4CREm-Luc was markedly decreased. Upon mutation of the CRE the promoter exhibited 29.18 ± 0.24% of BDNF4-Luc basal activity. The treatment with LiCl, 8-bromo-cAMP, or the combination of LiCl and 8-bromo-cAMP did not affect the promoter activity either under control conditions or upon overexpression of TORC1 (Figure 37). The statistical analysis by two-way ANOVA confirmed significant effects of lithium on cAMP-induced transcription mediated by BDNF4-Luc dependent on the overexpression of TORC1 with $p<0.0001$.

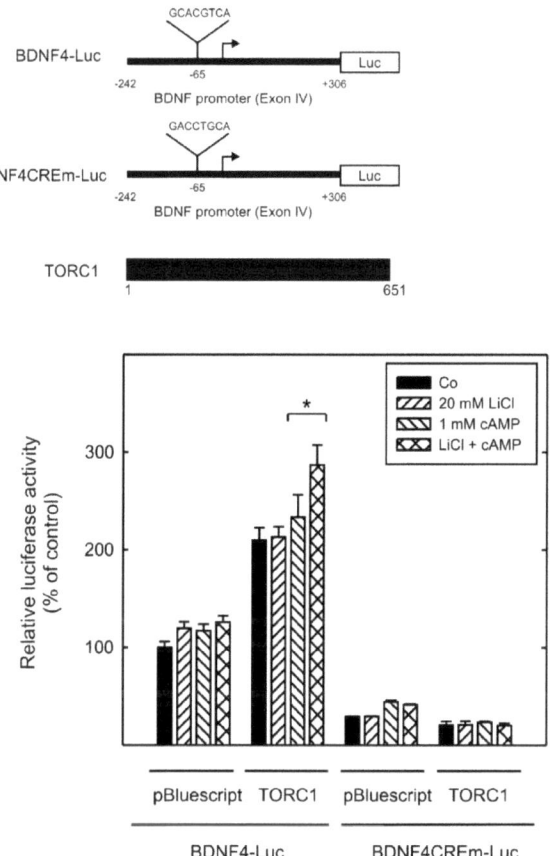

Figure 37: Effects of lithium and cAMP on at the transcriptional activity the human *BDNF(exonIV)*-gene promoter in luciferase reporter-gene assays.

HIT-T15 cells were transiently transfected with the luciferase-reporter gene BDNF4-Luc, comprising the promoter region -242 to +306 of the human *BDNF*-gene promoter of exon IV, or BNDF4CREm-Luc in which the CRE at position -65 was mutated. Cotransfected was an expression plasmid encoding full-length human TORC1. For control the empty expression vector pBluescript was cotransfected. A schematic illustration of the constructs is shown. The cells were treated with 20 mM LiCl for 7 h, with 1 mM 8-bromo-cAMP for 6 h, or with the combination of LiCl and 8-bromo-cAMP. Luciferase activity was determined. Relative luciferase activity values are means ± SEM of three independent experiments performed in duplicate, and are expressed in percent of basal activity of BDNF4-Luc.
Under control condition treatment with LiCl and 8-bromo-cAMP did not affect the promoter activity of BDNF4-Luc. Upon overexpression of TORC1, the basal promoter activity was increased. LiCl significantly increased the promoter activity in the presence of 8-bromo-cAMP. Mutation of the CRE decreased the promoter activity of BDNF4CREm-Luc to one third of BDNF-Luc. Statistical analysis was performed by two-way ANOVA followed by Student's *t*-test: *$p<0.05$.

17.c Effects of lithium at the human *NR4A2*-gene promoter

In luciferase reporter-gene assays, the effect of lithium on cAMP-induced transcription at the human *NR4A2*-gene promoter was investigated. HIT-T15 cells were transiently transfected with the luciferase-reporter gene NR4A2-Luc comprising the promoter region -389 to +154 of the human *NR4A2*-gene. To explore effects elicited by CREB cells were transfected with the luciferase-reporter gene NR4A2CREm-Luc in which the CRE at position -3 was destroyed by restriction digest, blunting and religation of the construct. Cotransfected was an expression plasmid encoding the full-length human TORC1 (Figure 38, top). For control the empty expression vector pBluescript was cotransfected. The cells were treated with 20 mM LiCl, with 1 mM 8-bromo-cAMP, or with the combination of both. Under control conditions the treatment with 20 mM LiCl did not change the basal promoter activity of NR4A2-Luc. The treatment with 1 mM 8-bromo-cAMP increased the promoter activity 1.3-fold to 127.89 ± 3.39% (p<0.001, Figure 38). LiCl increased the promoter activity induced by 8-bromo-cAMP 1.31-fold to 167.25 ± 6.51% (p<0.001, Figure 38). The overexpression of TORC1 increased the basal promoter activity of NR4A2-Luc 2.2-fold (p<0.0001). Treatment with LiCl did not affect the basal promoter activity of NR4A2-Luc, but 1 mM 8-bromo-cAMP increased the activity 1.88-fold to 415.12 ± 25.92% (p<0.0001) when TORC1 was overexpressed (Figure 38). 20 mM LiCl enhanced the 8-bromo-cAMP-induced promoter activity of NR4A2-Luc 1.69-fold to 700.75 ± 39.92% (p<0.001). The analysis by two-way ANOVA confirmed specific effects of lithium on cAMP-induced promoter activity of NR4A2-Luc with p<0.0001. The overexpression of TORC1 resulted in more pronounced effects with p<0.0001 as shown by two-way ANOVA. The mutation of the CRE in NR4A2CREm-Luc completely disrupted the promoter activity (Figure 38) compared to NR4A2-Luc (two-way ANOVA: p<0.0001). Both, treatment with 8-bromo-cAMP and LiCl or the overexpression of TORC1 did not enhance the promoter activity of NR4A2CREm-Luc (Figure 38).

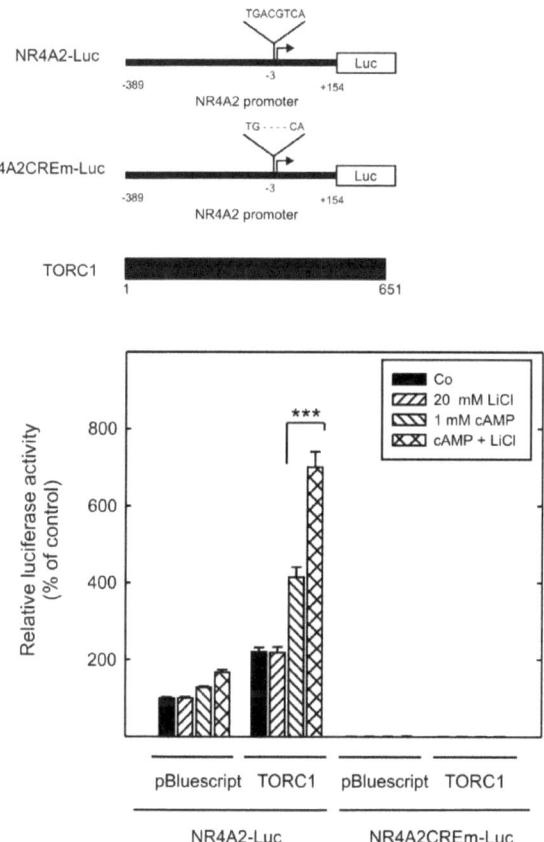

Figure 38: Effects of lithium and cAMP on human *NR4A2*-gene transcription in luciferase reporter-gene assays.

HIT-T15 cells were transiently transfected with the luciferase-reporter gene NR4A2-Luc, comprising the promoter region -389 to +154 of the human *NR4A2*-gene promoter, or NR4A2CREm-Luc in which the CRE at position -3 was destroyed. Cotransfected was an expression plasmid encoding full-length human TORC1. For control the empty expression vector pBluescript was cotransfected. A schematic illustration of the constructs is shown. The cells were treated with 20 mM LiCl for 7 h, with 1 mM 8-bromo-cAMP for 6 h, or with the combination of LiCl and 8-bromo-cAMP. Luciferase activity was determined. Relative luciferase activity values are means ± SEM of three independent experiments performed in duplicate, and are expressed in percent of basal activity of NR4A2-Luc. Statistical analysis was performed by two-way ANOVA followed by Student's t-test: ***$p<0.001$.

Under control condition treatment with 8-bromo-cAMP alone increased the promoter activity of NR4A2-Luc. LiCl increased the promoter activity induced by 8-bromo-cAMP. Upon overexpression of TORC1, the basal promoter activity was increased and the effects upon treatment were more pronounced. Deletion of the CRE abolished the promoter activity of NR4A2CREm-Luc. No effects of treatment where detectable under control conditions or upon overexpression of TORC1.

Discussion

In the present thesis the molecular mechanism was investigated by which lithium stimulates CRE/CREB-directed gene transcription through TORC in the presence of cAMP. Lithium did affect neither the nuclear translocation nor the intrinsic transcriptional activity of TORC proteins in HIT-T15 cells, but increased the oligomerization of TORC1 and the association of TORC with CREB. All three human TORC isoforms mediated the lithium-induced enhancement of cAMP-dependent CRE/CREB-directed transcriptional activity. As a cation, lithium facilitated directly the interaction between CREB and TORC1 in a concentration-dependent manner. In contrast to lithium, magnesium strongly inhibited the CREB-TORC1 interaction which was attenuated by lithium. The amino acid K290 of CREB is known to mediate the binding of a magnesium ion to the CREB bZip. In the present study, the role of CREB-K290 for the effect of lithium on the cAMP-induced CREB directed gene transcription was investigated. CREB-K290 mutants were inducible by lithium and lithium stimulated the interaction between CREB-K290 mutants and TORC1, whereas magnesium inhibited the interaction. In addition to its action on artificial promoters, lithium was shown in the present work to enhance the cAMP-induced gene transcription at CRE/CREB-dependent human native promoters of the *cfos-*, *BDNF(exonIV)-* and *NR4A2-*genes through TORC1.

1. HIT-T15 cells as a model system

TORC as a mediator of the effect of lithium on cAMP-induced CREB-directed transcription has been identified in the insulinoma tumor cell line HIT-T15 derived from hamster β-cells (Boer et al., 2007). HIT-T15 cells resemble neurons in important aspects: they express potassium channels and voltage-dependent calcium channels making them excitable by membrane depolarization (Schwaninger et al., 1993b). Moreover, they express proteins that are also typical for neurons such as the scaffold protein JNK-interacting protein (JIP)-1b / islet-brain (IB)1 (Hashimoto et al., 2002), and the transcription factors Pax6 (Cerf, 2006; D'Elia et al., 2006) and BETA2 / neuroD (Cerf, 2006; Peyton et al., 1996). Due to the similarities between pancreatic endocrine cells to neurons and endocrine cells originating from the ectoderm via the neural crest, it was thought initially that the pancreatic islet also originates from the ectoderm (Pearse and Polak, 1971). However, nowadays it is clear that the endocrine cells from the pancreas originate from the endoderm germ layer (Nekrep et al., 2008).

CREB-directed gene transcription in response to various stimuli, like cAMP analogs or potassium-induced membrane depolarization, has been well characterized in HIT-T15 cells (Boer et al., 2007; Oetjen et al., 1994; Oetjen et al., 2005; Schwaninger et al., 1995; Schwaninger et al., 1993a; Schwaninger et al., 1993b). Moreover, the translocation and function of TORC2 in response to cAMP and calcium has been examined before in HIT-T15 cells (Screaton et al., 2004). Therefore this cell line was used as a model of electrically excitable cells.

2. Concentrations of lithium used in the present study

In the present study, lithium at a concentration of 20 mM was used to enhance cAMP-induced CRE/CREB-directed gene transcription in HIT-T15 cells which was conferred by TORC1. Of note, the therapeutic serum levels of lithium in patients range from 0.5 to 1.2 mM (Mota de Freitas et al., 2006). However, by now it has not been determined at which concentration lithium is present in excitable cells in human brain tissue *in vivo* (Birch, 1999). From a study investigating the lithium transport in mouse brain using the stable isotopes ^{6}Li and ^{7}Li, one might speculate about an enrichment of lithium in brain tissue compared to the plasma concentration (Heurteaux et al., 1991). Heurteaux and colleagues presented mice with a 30 mM LiCl solution as beverage for 30 days. After 10 days the plasma concentration reached a plateau of 0.28 mM. Interestingly, the lithium concentration in different brain regions was increased to a total concentration of 1.8 mM in the neocortex, and concentrations in the hippocampus, the cerebellum (grey matter) and the thalamus were 1.6 mM, 1.7 mM, and 2.1 mM, respectively (Heurteaux et al., 1991). In HIT-T15 cells the intracellular concentration of lithium was determined to be 6-fold lower compared to the extracellular medium concentration (Boer et al., 2007). In that study, a dose-response curve revealed a significant enhancement of the cAMP-induced CRE/CREB-directed gene transcription by lithium already at extracellular concentrations of 6 mM lithium (Boer et al., 2007). Taking into account that the intracellular concentration was found to be 6-fold lower in HIT-T15 cells, lithium increases the cAMP-induced CRE/CREB-directed gene transcription at concentrations of 1 mM, hence within the therapeutic range.

3. Regulation of TORC by nuclear and cytosolic shuttling in HIT-T15 cells

Cyclic AMP- and Ca^{2+}-activated signaling pathways control the nucleo-cytoplasmic shuttling of TORC, a key regulator for CREB-directed transcriptional activity (Bittinger et al., 2004). In the present work, it was investigated whether the lithium-induced enhancement of CRE/CREB-directed gene transcription was due to increased nuclear translocation of endogenous TORC proteins. Under resting conditions TORC proteins were mainly located to the cytosol. The nuclear accumulation of TORC was induced in HIT-T15 cells by high potassium-induced membrane depolarization and blocked by the calcineurin inhibitor cyclosporin A consistent with the notion that calcineurin dephosphorylates TORC proteins in response to elevated calcium levels in HIT-T15 cells (Screaton et al., 2004). Furthermore, the nuclear translocation of endogenous TORC upon increased cAMP levels was evident in a concentration-dependent manner. These results confirm previous studies in which the nuclear accumulation of TORC in different cell lines and also primary hippocampal neurons was inducible by treatment with forskolin, a potent activator of AC (Bittinger et al., 2004; Kovacs et al., 2007; Screaton et al., 2004; Zhou et al., 2006). However, no effect of lithium on the translocation of TORC into the nucleus was detected in the present study, in presence or absence of cAMP. Thus an increased nuclear accumulation of TORC was ruled out to contribute to the effect of lithium on cAMP-induced CREB-directed transcription.

4. Lithium facilitates the oligomerization of TORC1

A highly conserved predicted coiled-coil structure is located at the N-terminus of TORC (Iourgenko et al., 2003). TORC proteins were found to form oligomers by this domain *in vitro* and were suggested to bind to the bZip of CREB as a tetramer (Conkright et al., 2003a). In the present study, the effect of lithium on the oligomerization of TORC was examined as a putative molecular target. Lithium was shown to facilitate the oligomerization of TORC1. Though the physiological relevance of the complex formation for the interaction with CREB has not been investigated so far, the increased oligomerization of TORC by lithium might play a role in the enhancement of CREB-directed gene transcription. One might speculate that the binding of lithium to TORC1 increases TORC1 oligomer stability, which might in turn stabilize the interaction between CREB and TORC (see below).

5. Lithium does not influence the transcriptional activity of TORC proteins

TORC proteins contain a strong transactivation domain in the C-terminal 200 amino acids, a glutamine-rich region found to associate with $TAF_{II}130$ of the basal transcriptional machinery (Conkright et al., 2003a; Iourgenko et al., 2003). This can be studied by the GAL4-system, where TORC is fused to the heterologous GAL4 DNA-binding domain and tested for induction of a promoter with GAL4-binding sites, as has been shown previously (Iourgenko et al., 2003). Consistent with reports by Iourgenko and colleagues, it was found in the present study that all three human TORC isoforms strongly induced the promoter (Iourgenko et al., 2003). Lithium did not affect the transcriptional activity of the GAL4-TORC proteins, alone or in combination with cAMP. However, the transactivation potential of TORC may not entirely be constitutive, since cAMP stimulated the transcriptional activity of GAL4-TORC1 and GAL4-TORC2. Using a related experimental design, the activity of the well-known CREB coactivator, namely CBP, is also stimulated by calcium and cAMP in HIT-T15 cells (Oetjen et al., 2005) as well as in the mouse pituitary cell line AT20 (Chawla et al., 1998), in rat primary cortical neurons (Hu et al., 1999) and rat primary hippocampal neurons (Impey et al., 2002). These data support the idea that CREB coactivators are targets of signalling cascades known to regulate the transcriptional activity of CREB. It remains unclear by which mechanism cAMP enhances the transcriptional activity of GAL4-TORC1 and GAL4-TORC2. Though it is generally assumed that GAL4-TORC proteins localize to the nucleus by the GAL4 NLS, the cytosolic-nuclear shuttling of GAL4-TORC proteins might be involved here. The failure of cAMP to increase transcriptional activity of GAL4-TORC3 would then be consistent with reports in which TORC3 was localized to the nucleus also under basal conditions in certain cell lines (Bittinger et al., 2004; Katoh et al., 2006; Screaton et al., 2004). An alternative mechanism is offered by NONO (non-POU-domain-containing octamer-binding protein) as cAMP has been shown to stimulate the binding of TORC to NONO, which then acts as a bridge between the CREB/TORC complex and RNA polymerase II (Amelio et al., 2007). Altogether these results indicate that lithium increases neither nuclear translocation of TORC nor TORC transcriptional activity to stimulate cAMP-induced CREB-directed transcription.

6. Lithium facilitates the interaction between CREB and TORC proteins

That TORC confers the enhancement by lithium of the cAMP-induced CREB-directed transcription becomes apparent when TORC1 is expressed together with the transcriptionally inactive bZip fused to the DNA-binding domain of GAL4. In this context, TORC1 restores the transcriptional activity in response to cAMP and the enhancement by lithium (Boer et al., 2007). In the present study, all three human isoforms of TORC were shown to restore the transcriptional activity of the GAL4-bZip and, moreover, to be able to mediate the enhancement by lithium of cAMP-induced transcription. Interestingly, TORC2 and TORC3 are 32% homologue to TORC1, but the three proteins are highly conserved in the first 50 amino acids, the domain mediating the interaction with CREB (Conkright et al., 2003a; Iourgenko et al., 2003). Due to this high similarity in the N-terminal CREB-interaction domain, the effect of lithium on the interaction between CREB and the three TORC isoforms was investigated. Lithium strongly increased the interaction between the bZip and all three TORC isoforms in a cellular mammalian two-hybrid assay using only the interaction domains of the proteins. In contrast, lithium increased the interaction only between CREB and TORC1 in the cell-free GST pull-down assay. One apparent difference between the two-hybrid assay and the GST pull-down assay is the usage of only the interaction domains versus the full-length proteins, respectively. Therefore, the GST pull-down assay was repeated with truncated TORC proteins comprising only the N-terminal half. As with the full-length TORC proteins, lithium again increased the interaction only between CREB and TORC1. Thus, it is likely that the reason for these diverging results might be the absence of the cellular context and posttranslational modifications in the GST-pull down-assay. When taken together, the results of the present study indicate that lithium facilitates the interaction between CREB and TORC proteins, although posttranslational modifications seem to be required for TORC2 and TORC3.

In this study, TORC1 was identified to be the predominantly expressed TORC isoform in HIT-T15 cells. Importantly, this isoform is mainly found in brain tissue and neurons and was implicated to be involved in long-term potentiation (Conkright et al., 2003a; Kovacs et al., 2007; Zhou et al., 2006). Due to these facts, the following investigations focussed on this isoform.

Consistent with a previous report (Conkright et al., 2003a), the concomitant binding of CREB and TORC1 to the CRE was detectable in the EMSA. To further characterize the effects of lithium on the interaction between CREB and TORC1, the recruitment of TORC1

Discussion

to the promoter was investigated by chromatin immunoprecipitation. In the cellular context, lithium increased the amount of DNA precipitated by TORC1 in the presence of cAMP, indicating enhanced promoter occupancy by TORC1 through binding to CREB. TORC1 on its own does not contain a DNA-binding domain (Iourgenko et al., 2003). Lithium alone or cAMP alone did not detectably affect the promoter occupancy of TORC1. Of note, an increase has been shown before for TORC2 upon forskolin stimulation (Koo et al., 2005; Ravnskjaer et al., 2007). The system used in the present study might be not sensitive enough to detect the effect elicited by the treatment with 8-bromo-cAMP alone. In fact, stimulation with 10 µM forskolin is in general about 10-fold more effective compared to 2 mM 8-bromo-cAMP (Oetjen et al., 1994; Schwaninger et al., 1995). Noteworthy, lithium stimulates TORC-CREB binding independently of cAMP and protein kinase A activity, since in the mammalian two-hybrid assay lithium induced a robust increase in the binding of the TORC N-terminus to the CREB bZip in the absence of cAMP and, in addition, cAMP treatment had no effect on the interaction. Furthermore, lithium increased the interaction between CREB and TORC1 in a concentration-dependent manner in the GST pull-down assay, as it did enhance the cAMP-induced CREB-directed transcription in cell culture (Boer et al., 2007). The CREB mutant R300A did not show significant interaction with TORC1 under cell-free conditions as well as in the cellular system confirming the results of Screaton and colleagues who demonstrated the disruption of the CREB-TORC interaction upon mutation of R300 (Screaton et al., 2004). Strikingly, a TORC1 construct lacking the first 44 amino acids did not confer any transcriptional activity to the GAL4-bZip. These data further support the view that only the first 44 amino acids of TORC1 are necessary and sufficient to mediate the interaction with bZip and the enhancement of the interaction by lithium.

Taken together, these results support the view that, lithium stimulates cAMP-induced CREB transcriptional activity by enhancing the binding of TORC to CREB once TORC has been shifted into the nucleus by cAMP.

7. Magnesium inhibits the interaction between CREB and TORC1

The similarity of lithium to magnesium allows lithium to interfere with magnesium-dependent processes. Actually, substitution of lithium for magnesium in certain enzymes is a known mechanism by which lithium influences signalling cascades involving GSK3β and adenylyl cyclase (Birch, 1999; Mota de Freitas et al., 2006; Quiroz et al., 2004). Therefore, the effect of magnesium on the CREB-TORC1 interaction was examined. In contrast to lithium, magnesium strongly inhibited the interaction between CREB and TORC1 in a concentration-dependent manner. This effect was antagonized by lithium, which increased the TORC-CREB binding by relief from magnesium inhibition. These data implicate a common but reciprocal influence of lithium and magnesium on the interaction between CREB and TORC1. Interestingly, the crystal structure of the CREB bZip revealed the presence of a hexahydrated magnesium ion in the cavity between the bifurcating basic regions (Schumacher et al., 2000). This magnesium ion is coordinated by the extended side chains of the amino acid residue lysine at position 290 (K290) (Craig et al., 2001). The functional importance of this lysine residue was explored already in 1990 by Dwarki and colleagues who reported the abolishment of CREB-DNA binding and therefore loss of function when K290 was mutated to glutamate (K290E) (Dwarki et al., 1990). Similarly, Craig and coworkers mutated K290 to alanine (K290A) and pinpointed the loss of CREB-DNA binding to the disruption of the magnesium-binding properties whereas the overall protein-stability was not affected (Craig et al., 2001).

8. The role of the CREB K290 mutation for the effect of lithium on CREB-TORC1 interaction

To examine the role of the K290 of CREB for the enhancement of lithium on cAMP-induced CREB-directed transcription this residue was substituted with glutamic acid (K290E) or alanine (K290A). Consistent with the results from Dwarki and coworkers (Dwarki et al., 1990), CREB-K290E showed in the EMSA a markedly reduced affinity for the CRE of the rat *somatostatin* gene promoter.

To characterize these mutants with respect to their transcriptional activity, the GAL4-system was used, allowing the examination of a transcription factor independent of its DNA-binding ability. After expression of GAL4-CREB, both mutations of CREB-K290 (K290E, K290A) increased the transcriptional activity of CREB both under basal conditions and in the presence of cAMP. Despite elevated transcriptional activities of the mutant proteins, lithium further enhanced cAMP-dependent transactivation by GAL4-CREB-K290E and GAL4-CREB-K290A. Also, the stimulation of transcription by TORC1 through GAL4-CREB-bZip was increased by both mutations of K290 within CREB bZip under basal conditions as well as in the presence of cAMP, and the TORC1-dependent transcriptional activity of both mutants in the presence of cAMP was further enhanced by lithium. In control experiments the stimulation of transcription by TORC1 was not observed when only the isolated DNA-binding domain of GAL4 was coexpressed, indicating that the TORC1-dependent transcriptional activity of the GAL4-CREB proteins is mediated by a specific interaction between TORC and the CREB-bZip domain. In addition, the elevated activities of the mutant GAL4-CREB proteins (K290E, K290A) are not due to increased expression levels as shown by Western blotting. Therefore, when taken together, these results indicate that the mutation of K290 within the CREB bZip into glutamic acid or alanine increases the TORC1-dependent CREB transcriptional activity; furthermore, these mutations did not prevent the stimulation by lithium of cAMP-induced TORC1-dependent CREB activity. Under cell-free conditions, the mutation of K290 to glutamic acid (GST-CREB-K290E) or alanine (GST-CREB-K290A) decreased the binding of [^{35}S]TORC1 to CREB but did not prevent the stimulation by lithium of [^{35}S]TORC1 binding. However, though lithium stimulated the interaction between TORC1$_{1-44}$ and bZip-K290E or bZip-K290A in the mammalian two-hybrid assays, the effect of lithium was significantly lower compared to bZip wild-type. This result did not depend on different expression levels of the proteins because an equal expression was assured beforehand. To examine the effect of CREB-K290 mutations on the recruitment of TORC by CREB *in vivo*, chromatin immunoprecipitation was performed. Promoter occupancy of TORC1 in the presence of cAMP was increased by both K290 mutations. This suggests that mutating K290 enhances the binding of TORC1 to the CREB bZip. After expression of GAL4-bZip wild-type and in the presence of cAMP, lithium stimulated TORC1 promoter occupancy. Both mutations of K290 did not prevent the stimulation by lithium of TORC1 recruitment by CREB bZip. Though opposing the results from the cell-free GST pull-down assay, the chromatin immunoprecipitation assay indicates an increased interaction between CREB K290-mutants and TORC1 at the promoter which is accordable with an increased basal activity of the mutants. A more intense interaction provides also an explanation for the increased responsiveness to cAMP stimulation of the K290-mutants.

Thus, the mutation of K290 to glutamic acid or alanine prevented neither the stimulation by lithium of TORC binding to CREB *in vitro* (GST pull-down assay) and *in vivo* (chromatin immunoprecipitation) nor the lithium induced increase in the transcriptional activity. Finally, the mutation of K290 did not prevent the inhibition by magnesium of TORC binding to CREB *in vitro*, suggesting a site of magnesium action distinct from K290. Because magnesium ions have to bind with relatively high affinity to their binging sites to be incorporated in a crystal during the process of crytallization (Mota de Freitas et al., 2006), a low affinity magnesium binding site may thus exist besides K290 and be more susceptible to lithium/magnesium competition. This has been shown for the G-protein α subunit $G_{i\alpha 1}$. In x-ray crystallographic studies, only one magnesium ion was found coordinated to $G_{i\alpha 1}$ (Coleman and Sprang, 1998). The monitoring of its intrinsic tryptophan fluorescence, however, showed two distinct magnesium binding sites and indicated that lithium may compete for only the low affinity magnesium binding site (Minadeo et al., 2001; Mota de Freitas et al., 2006). Therefore, while the results of the present study may exclude K290 and may rather suggest another, low-affinity magnesium binding pocket, the lithium binding site on the CREB-TORC complex remains to be defined through which lithium stimulates TORC association and CREB activity.

Whereas it may not mediate lithium action, the results of the present study suggest that CREB-K290 is, nevertheless, involved in the interaction between TORC and CREB. The only amino acid so far known to be involved in TORC binding is the CREB leucine zipper residue R300 (Screaton et al., 2004). Because the TORC-CREB association is disrupted by the addition of 400 mM KCl and may thus be driven by electrostatic interactions (Conkright et al., 2003a), it has been suggested that R300 may be required for TORC binding by providing an attractive positive charge (Screaton et al., 2004). However, R300 occupies the g position of the first heptad repeat of the CREB leucine zipper and, as shown in the crystal structure, undergoes ionic interactions with E305 of the other subunit (Schumacher et al., 2000; Vinson et al., 2002). The mutation of R300 is thus expected to decrease dimer stability (Schumacher et al., 2000; Vinson et al., 2002), which may harm TORC binding. The results of the present study now suggest the involvement of a second amino acid in TORC binding, the CREB basic region residue K290, thereby indicating that the CREB contact surface for the interaction with TORC extends from the leucine zipper to the basic region. This may be important, because contact surfaces are generally thought to allow better regulation of the interaction (Vinson et al., 2002). The mutation of K290 increased the association of TORC with CREB *in vivo* as revealed by chromatin immunoprecipitation. Inappropriate folding of the bacterially expressed proteins, missing posttranslational modifications and cellular proteins may explain the contrasting effect of the mutation *in vitro*. Consistent with and confirming the increase in TORC binding *in vivo*, the mutation of K290 also increased CREB transcriptional activity and the activity conferred by TORC to the CREB bZip. Since at the same time the mutation of K290 virtually abolished CREB binding to the consensus CRE as has been shown before (Craig et al., 2001; Dwarki et al., 1990), the data concur with the previous conclusion that the association of TORC with CREB does not require DNA binding (Conkright et al., 2003a). The basic residue K290 was mutated to two very different amino acids, to the acidic residue glutamic acid and the neutral residue alanine. The fact that both substitutions produced the same effect renders it very likely that the mutations enhanced TORC binding and CREB transcriptional activity because K290 imposes a burden on the interaction between TORC and CREB. At this point it is only possible to speculate about the reason of K290 repulsion for TORC-CREB complex formation. In other cases of protein-protein binding, it is quite common to find repulsive interactions in addition to attractive ones and they are thought to be important to establish the specificity of binding (Vinson et al., 2002). Furthermore, K290 has most likely due to proton transfer, an electrostatic interaction with a coordination water of the magnesium cation (Schumacher et al., 2000) and, in the absence of magnesium, is able to interact with the phosphate backbone of the DNA (Craig et al., 2001). If this or another interaction would reposition the lysine side chain or bury charges, it may reduce the hindrance by K290 and thus favor TORC-CREB binding. In this way, K290 may allow the regulation of the association of TORC with CREB. Thus, the results of the present study clearly indicate that CREB-K290 not only mediates DNA binding but is also involved in the association of TORC with CREB and thus the implementation of CREB transcriptional activity.

9. Lithium enhances the cAMP-induced CREB-directed gene transcription at native human promoters

To further explore the physiological significance of the described effect of lithium on cAMP-induced CREB-directed gene transcription mediated by TORC1, three CRE/CREB-directed native gene promoters were examined. Depending on TORC1 overexpression, lithium enhanced the cAMP-induced CREB-directed transcription for the *fos-*, *BDNF(exon IV)-*, and *NR4A2*-gene promoters.

9.a The human *fos*-gene promoter

The *fos*-gene is a proto-oncogene encoding cFos, a component of activator protein 1 (AP1) transcription factor complexes. It is an immediate-early gene induced by various stimuli and is widely used as anatomical marker for activated neurons in the central nervous system (Kovacs, 2008). cFos forms heterodimers with cJun to build the AP1 heterodimeric transcription factor belonging to the family of bZip transcription factors (van Dam and Castellazzi, 2001). In the present study the promoter of the *fos*-gene was employed since it is a well described target of CREB-directed gene transcription inducible by cAMP and calcium (Sassone-Corsi et al., 1988). It was demonstrated in the present study that lithium enhanced the cAMP-induced *fos* gene transcription when TORC1 was overexpressed. Though the basal promoter activity was markedly reduced, the mutation of the CRE did not block this effect. Of note, the *fos*-gene promoter contains also an AP1 consensus site. Recently, TORC1 was shown to interact also with the AP1 heterodimer made up of cFos and cJun and to promote AP1-directed transcription in response to phorbol ester (Canettieri et al., 2009). However, AP1-directed transcription was not increased in response to cAMP (Canettieri et al., 2009), and is in general found to be activated by growth factors and phorbol ester instead of cAMP (Kovacs, 2008). This result argues against an involvement of AP1 in the present effect of lithium on cAMP-induced transcription observed upon mutation of the CRE in the *fos*-gene promoter. Rather the similarity of the AP1 consensus site to the CRE site might play a role. The AP1 core sequence 5'-TGAC/GTCA-3' differs only by one base from the ideal CRE core sequence 5'-TGACGTCA-3' (van Dam and Castellazzi, 2001). Importantly, the central CG of the CRE is thought to be necessary for high affinity binding of CREB (Deutsch et al., 1988; Schumacher et al., 2000). But the binding of CREB to the AP1 site has been reported as well, though binding occurs with much lower affinity (Kerppola and Curran, 1993). The sequence of the AP1 site of the *fos*-gene promoter used in the present study is 5'-TGCGTCA-3' containing the denoted necessary central CG (Konig et al., 1989). In addition, a study reported the predominant binding of CREB to one of the four AP1 sites in the *NR4A1*-gene promoter. The functional examination revealed also responsiveness to cAMP indicating the involvement of CREB (Inaoka et al., 2008). The sequence of the AP1 site in the *NR4A1*-gene promoter is 5'-TGCGTCA-3' (Inaoka et al., 2008), identical to that of the *fos*-gene promoter used in the present study. These data support the idea that the binding of CREB to the AP1 site might compensate for the mutation of the CRE of the *fos*-gene promoter, thereby providing an explanation for the observed effects.

9.b The human *BDNF(exon IV)*-gene promoter

The *BDNF*-gene encodes the brain-derived neurotrophic factor (BDNF) belonging to the family of nerve growth factors. BDNF acts on the tyrosine kinase-coupled receptor B (TrkB). Thereby it exerts numerous intracellular effects via mitogen-activated protein kinase (MAPK), phosphatidylinositol 3-kinase (PI3-K), and phospholipase C (PLC) α signal transduction pathways. BDNF has a well-established role in the development, survival and differentiation of select neuron populations and is implicated in long-term potentiation (Nair and Vaidya, 2006; Post, 2007). The human *BDNF* gene is composed of eleven exons and contains nine functional promoters leading to multiple transcripts with different expression pattern. Exon IX forms the coding exon (Pruunsild et al., 2007). In the present study, the promoter of exon IV of the human BDNF gene was employed. BDNF transcripts containing exon IV are highly expressed in adult human brain tissue in regions involving the cerebellum, amygdala, hippocampus, and frontal cortex (Pruunsild et al., 2007), regions reported to show volumetric abnormalities in patients with BD (Frey et al., 2007; Scherk et al., 2004). CREB was reported to bind to the promoter of exon IV of the human *BDNF* gene. In addition, this is so far the only *BDNF* promoter reported to be responsive to cAMP (Fang et al., 2003). In contrast to the findings from Fang and colleagues (2003), we did not observe increased transcriptional activity in response to cAMP without the overexpression of TORC1. However, lithium increased the cAMP-induced transcription in the presence of TORC1. The mutation of the CRE disrupted the effect of lithium, underlining the necessity of CREB for this effect to occur.

Noteworthy, the promoter of exon IV of the human *BDNF* gene does not contain a TATA-box proximal to the putative transcription start site (Fang et al., 2003). Conkright and coworkers identified the requirement of a proximal TATA-box for the cAMP-inducibility of CRE/CREB-directed gene transcription. TATA-containing promoters showed a strong induction by cAMP-analogs whereas TATA-less promoters were only marginally inducible. The insertion of a TATA-box in a TATA-less promoter restored the cAMP-inducibility (Conkright et al., 2003b). Furthermore, the potentiation of CREB-directed gene transcription by TORC seems to require the presence of a consensus TATA-box, too. Compared to TATA-containing promoters the increase of promoter activity upon overexpression of TORC was markedly reduced for TATA-less promoters (Conkright et al., 2003a). Thus, the absence of a consensus TATA-box in the promoter of the exon IV of the human BDNF gene might account for the not detectable cAMP responsiveness, even upon overexpression of TORC1. However, a significant enhancement by lithium was observed. Noteworthy, BDNF has been suggested to be involved in recurrent mood disorders. For instance, low BDNF serum levels have been reported for patients with BD in the depressive as well as in the manic phase (Hashimoto et al., 2004; Post, 2007). Long-term lithium application was reported to increase levels of BDNF expression in the frontal cortex and hippocampus in rats (Einat et al., 2003; Fukumoto et al., 2001; Omata et al., 2008). Moreover, the expression of BDNF from a CREB-dependent promoter was strongly induced by cAMP and calcium in mouse cortical neurons, whereas the calcineurin inhibitor cyclosporin A abolished this effect (Kovacs et al., 2007), arguing for the involvement of TORC1 in *BDNF*-gene expression. In view of these findings, one may assume that lithium acting at the *BDNF(exonIV)* promoter contributes to the therapeutic actions of the drug.

9.c The human *NR4A2*-gene promoter

The *NR4A2* gene is a known CREB target gene and encodes the orphan nuclear receptor Nurr1, also referred to as NR4A2, NOT, RNR-1, or HZF-3 (Conkright et al., 2003b; Perlmann and Wallen-Mackenzie, 2004). Nurr1 is structurally related to members of the nuclear receptor family of transcription factors which are ligand-regulated. Although Nurr1 contains a ligand-binding domain which is folded almost in the same way as in the other nuclear receptors, the space normally occupied by ligands in nuclear receptor is entirely filled by hydrophobic amino acid side chains, as revealed by X-ray crystallography of Nurr1 (Maxwell and Muscat, 2006; Perlmann and Wallen-Mackenzie, 2004). Thus, Nurr1 on its own lacks ligand-binding capacity. On the other hand, Nurr1 like other nuclear receptors forms heterodimers with RXR, another nuclear receptor. RXR regulates transcription in a ligand-dependent manner and Nurr1/RXR heterodimers can be activated by RXR-ligands (Perlmann and Wallen-Mackenzie, 2004). Nurr1 is highly expressed in developing and mature brain tissue including cortex and hippocampus and is strongly implicated in the differentiation and maintenance of dopaminergic neurons in the central nervous system (Perlmann and Wallen-Mackenzie, 2004). The *NR4A2* gene belongs to the immediate early genes and can be activated by a variety of stimuli such as neurotransmitters, fatty acids, prostaglandins, growth factors, calcium, inflammatory cytokines, and membrane depolarization (Maxwell and Muscat, 2006). It has previously been shown that TORC expression activates the CRE/CREB-dependent *NR4A2* gene transcription and that TORC associates with the *NR4A2*-gene promoter upon increased levels of cAMP and calcium (Conkright et al., 2003a; Screaton et al., 2004). In the present study, lithium strongly increased cAMP-induced *NR4A2* gene transcription when TORC1 was overexpressed. This may be mediated by CREB as the destruction of the CRE by deletion of 4 bases completely eliminated the transcriptional activity. The effect of this CRE mutation was very marked for both cAMP-induced and basal *NR4A2* promoter activity. This might be due to the fact that the CRE of the *NR4A2* promoter is very close to the putative transcription start site (Torii et al., 1999), thus deletion of bases might prohibit transcription at all.

Nurr1 may play a role in cognitive processes because in rats Nurr1 mRNA levels were increased after spatial discrimination learning tasks (Pena de Ortiz et al., 2000). Moreover, the knockdown of Nurr1 in the hippocampus of rats by intrahippocampal infusion with Nurr1 antisense oligodeoxynucleotides reduced the spatial discrimination learning and memory, implicating Nurr1 in long-term memory and hippocampal-dependent cognitive processes (Colon-Cesario et al., 2006). Thus not only CREB but also Nurr1, whose expression is regulated by CREB, seems to be involved in long-term changes in the hippocampus. Furthermore, post-mortem tissue examination of patients with bipolar disorder revealed reduced mRNA levels of Nurr1 in the prefrontal cortex (Xing et al., 2006), implicating the association of Nurr1 with BD. The transcription of the *BDNF* gene also was shown to be Nurr1-dependent in rat primary neurons from the ventral midbrain, implicating further neuroprotective functions of Nurr1 (Volpicelli et al., 2007). In this context, lithium treatment might have beneficial effects via two ways. On the one hand, lithium activates directly Nurr1 expression by increasing the cAMP-induced CREB-directed transcription of the *NR4A2* gene, thereby putatively improving cognitive function. On the other hand increased Nurr1 levels might activate BDNF expression.

10. CREB and TORC1 – functional implications to neuroplasticity

Processes of neuroplasticity refers to the ability of the brain to adapt and change in response to internal and external stimuli. It is thought to underlie all types of memory formation and learning including reward- and stress-related responses (McClung and Nestler, 2008). The cellular mechanisms most likely involved in long-lasting changes in strength of neuronal connection in the brain are long-term potentiation (LTP) and long-term depression (LTD). The induction of LTP is generally believed to require synaptic activation of glutamate receptors of the NMDA (N-methyl-D-aspartate) type during postsynaptic depolarization mediated by glutamate receptors of the AMPA (α-amino-3-hydroxy-5-methyl-4-isoxazolepropionic acid) type (Malenka and Bear, 2004; Miyamoto, 2006). Activation of NMDA receptors leads to an influx of calcium activating the calcium/calmodulin-dependent kinase II (CaMKII). CaMKII is the best-described mediator of LTP, but others were implicated as well, for instance PKC or the mitogen-activated protein kinase (MAPK) cascade which activates extracellular signal-regulated kinases (ERKs) (Malenka and Bear, 2004). LTP can be devided into two phases, a short initial phase and a long-lasting phase. Initially, an increase in synaptic strength occurs upon activation of NMDA receptors lasting 30 to 60 min. The underlying mechanism is not completely understood, but so far it is clear that a major component of the initial phase of LTP is an increase in the number of AMPA receptors in the synaptic plasma membrane via activity-dependent changes in AMPA receptor trafficking. Moreover, the biophysical properties of the AMPA receptor are modified by phosphorylation (Malenka and Bear, 2004). Thus the short phase involves the cellular distribution and modification of pre-existing proteins. On the other hand, the late-phase or maintenance of LTP can last for hours or even days and requires gene transcription and *de novo* protein synthesis (Malenka and Bear, 2004; Wu et al., 2007).

CREB has long been implicated in LTP. Molecules thought to convey the activity induced by LTP to the nucleus are PKA, CaMK IV and MAPK which in turn activate CREB. In this context, calcium is thought to activate a calcium/calmodulin-sensitive AC triggering the production of cAMP and thereby activating PKA (Lonze and Ginty, 2002). CREB activity is often correlated with its status of phosphorylation and it was shown that activation of NMDA receptor increased the phosphorylation of CREB at S119 (Malenka and Bear, 2004). Supporting the notion that CREB is necessary for long-term changes, mice lacking the α-CREB and Δ-CREB isoforms showed impaired memory and deficits in hippocampal

LTP (Lonze and Ginty, 2002; Miyamoto, 2006). In addition, the overexpression of a constitutive active form of CREB in the CA1 region of the hippocampus lowers the threshold to induce persistent late phase LTP (McClung and Nestler, 2008). Recently, TORC1 was identified to potentiate CREB-directed transcription in response to cAMP and calcium stimuli in primary hippocampal neurons of rat and mice (Kovacs et al., 2007; Zhou et al., 2006). Noteworthy, TORC1 was reported to be involved in the late phase of LTP induced by tetanic stimulation of acute hippocampal slices of rat brain. The overexpression of a dominant negative form of TORC1 significantly reduced the potentiation induced by tetanic stimulation (Kovacs et al., 2007). Moreover, the nuclear accumulation of TORC1 correlated with induction of late phase LTP by a typical stimulation protocol whereas phosphorylation of CREB at S119 did not (Zhou et al., 2006). The present study shows that lithium enhances the cAMP-induced CREB-directed transcriptional activity by an increased interaction between CREB and TORC1. Therefore, one may assume that lithium contributes also to late phase LTP mediated by CREB. This new mechanism of lithium action thus could contribute to a form of drug-induced therapeutically relevant neuroplasticity.

11. Lithium, bipolar disorder and neuroplasticity

The changes in brain volume observed in patients with bipolar disrorder were mainly found to be due to reductions in glial cell density in the respective areas (Manji et al., 2003; Rajkowska, 2002). Importantly, glial cells are involved in the regulation of extrasynaptic glutamate concentration (Manji et al., 2003). Overstimulation of NMDA receptors with glutamate can induce excitotoxic cell death by an extensive increase in intracellular calcium (Hartnett et al., 1997). In addition, magnesium ions probably enter the cell as well through the activated ion channel of the NMDA receptor (Stout et al., 1996). Besides increasing calcium, magnesium also contributes to glutamate-induced excitotoxicity mediated by extrasynaptic NMDA receptors (Hartnett et al., 1997). The intracellular concentration of unbound magnesium in a mammalian cell is 0.2 to 1.2 mM. However, this represents only less than 10 % of the total magnesium (Flatman, 1991; Traut, 1994). In primary cortical neuron, for example, the intracellular concentration of free magnesium can be raised up to 10 mM and higher upon glutamate stimulation of NDMA receptors (Brocard et al., 1993; Hartnett et al., 1997). In view of the results of the present study, one might speculate that such fluxes in the concentration of magnesium ions could disrupt

TORC-CREB interaction. Lithium may reverse this effect and increase the binding of TORC to CREB, thus restoring CREB activity and the expression of CREB target genes like BDNF and Bcl-2 which promote cell survival (Chuang et al., 2002).

Consistent with the notion that the volume reductions in the brain of patient with BD are illness related (Frey et al., 2007; Scherk et al., 2004), lithium treatment was reported to increase the volume of the affected brain areas. This effect was observed for diseased subjects but not in healthy volunteers (Foland et al., 2008; Scherk et al., 2004; Yucel et al., 2007; Yucel et al., 2008). Brain areas involved in regulation of mood and behaviour, such as prefrontal cortex, the limbic system and the cerebellum display reduced volume in patients with BD and are also tissues highly expressing TORC1, arguing for the physiological relevance of the presently described effect of lithium on the interaction between CREB and TORC1. Thus the stimulation by lithium of the binding of TORC to CREB, as demonstrated by the present study, represents a novel mechanism of lithium action (Figure 39) that may contribute to the clinical mood stabilizing effect of lithium salts.

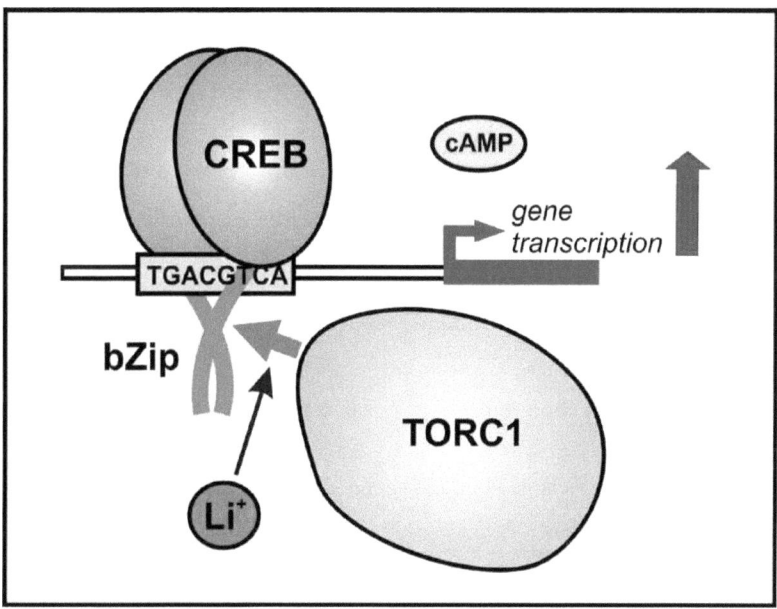

Figure 39: Novel mechanism of lithium action.
The present study provided evidence that lithium directly facilitates the interaction between CREB and TORC1 at the bZip thereby enhancing cAMP-induced CRE/CREB-directed gene transcription. This represents a new mechanism of lithium action which may contribute to the clinical efficacy of lithium treatment in bipolar disorder.

Summary and conclusion

Lithium salts are used now for 60 years to treat bipolar disorder, a chronic, severe and often life-threatening illness characterized by recurrent phases of the two opposing mood states of depression and mania. The mechanisms accounting for the clinical efficacy are not completely understood. Chronic treatment with lithium is required to establish mood stabilization, suggesting the involvement of processes of neural plasticity. CREB (cAMP response element binding protein) is a transcription factor known to mediate neuronal adaptation. Recently, the CREB-coactivator TORC (transducer of regulated CREB) has been identified as a novel target of lithium and was shown to confer an enhancement by lithium of cAMP-induced CREB-directed gene transcription. TORC is sequestered in the cytoplasm and its nuclear translocation controls CREB activity by interacting with the basic leucine zipper of CREB. In the present study, the molecular mechanism was investigated by which lithium stimulates cAMP-induced CRE/CREB-mediated gene transcription through TORC1.

For this purpose, the effect of lithium on the nuclear localization of TORC was investigated in HIT-T15 cells by immunofluorescence and the effect on TORC transcripitonal activity was examined by luciferase reporter-gene assays. The interaction between TORC and CREB was investigated *in vitro* in a GST pull-down assay, *in vivo* in a mammalian two hybrid assay and at the promoter by chromatin immunoprecipiation. Lithium did affect neither the nuclear translocation nor the intrinsic transcriptional activity of TORC proteins in HIT-T15 cells, but increased the oligomerization of TORC1 and the association of TORC with CREB. When heterologously expressed in HIT-T15 cells, all three human TORC isoforms mediated the lithium-induced enhancement of cAMP-dependent CRE/CREB-directed transcriptional activity. TORC1 was identified to be isoform predominantly expressed endogenously in HIT-T15 cells. As a cation, lithium facilitated directly the interaction between CREB and TORC1 in a concentration-dependent manner. The data support the notion that, once TORC has been shifted by cAMP into the nucleus, lithium facilitates the interaction between the CREB bZip and TORC1, thereby enhancing CREB-directed transcription. In contrast to lithium, magnesium strongly inhibited the CREB-TORC1 interaction which was attenuated by lithium. This effect could be considered as therapeutically relevant for the neuroprotective action of lithium possibly in that lithium counteracts excitotoxic effects of magnesium. The amino acid K290 of CREB, known to mediate the binding of a magnesium ion to the CREB bZip, appears not to be required for

Summary and conclusion

the effect of lithium as CREB-K290 mutants were inducible by lithium. In addition to its action on artificial promoters, lithium was shown to enhance the cAMP-induced gene transcription at CRE/CREB-dependent human native promoters of the *cfos-*, *BDNF(exonIV)-* and *NR4A2-*genes through TORC1. Thus, a physiological relevance of the presently described effect is suggested and implicates a contribution of lithium to neuroplasticity. Taken together the results of the present study provide a novel mechanism of lithium action that may contribute to the therapeutic effect of lithium in the treatment of bipolar disorder.

References

Adler, C.M., DelBello, M.P. and Strakowski, S.M. (2006) Brain network dysfunction in bipolar disorder. *CNS Spectr*, **11**, 312-320; quiz 323-314.

Amelio, A.L., Miraglia, L.J., Conkright, J.J., Mercer, B.A., Batalov, S., Cavett, V., Orth, A.P., Busby, J., Hogenesch, J.B. and Conkright, M.D. (2007) A coactivator trap identifies NONO (p54nrb) as a component of the cAMP-signaling pathway. *Proc Natl Acad Sci U S A*, **104**, 20314-20319.

Baillieux, H., De Smet, H.J., Paquier, P.F., De Deyn, P.P. and Marien, P. (2008) Cerebellar neurocognition: insights into the bottom of the brain. *Clin Neurol Neurosurg*, **110**, 763-773.

Baldacara, L., Borgio, J.G., de Lacerda, A.L. and Jackowski, A.P. (2008) Cerebellum and psychiatric disorders. *Rev Bras Psiquiatr*, **30**, 281-289.

Baldessarini, R.J. and Tarazi F.I. (2006) Pharmacotherapy of psychosis and mania. *Chapter 18 in Goodman & Gillman's The Pharmacological Basis of Therapeutics, 11thEdition*, 461-500. McGraw-Hill Medical Publishing Division, MacGraw-Hill, New York,USA

Belmaker, R.H. (2004) Bipolar disorder. *N Engl J Med*, **351**, 476-486.

Bhat, R.V., Budd Haeberlein, S.L. and Avila, J. (2004) Glycogen synthase kinase 3: a drug target for CNS therapies. *J Neurochem*, **89**, 1313-1317.

Birch, N.J. (1999) Inorganic pharmacology of lithium. *Chem Rev*, **99**, 2659-2682.

Bittinger, M.A., McWhinnie, E., Meltzer, J., Iourgenko, V., Latario, B., Liu, X., Chen, C.H., Song, C., Garza, D. and Labow, M. (2004) Activation of cAMP response element-mediated gene expression by regulated nuclear transport of TORC proteins. *Curr Biol*, **14**, 2156-2161.

Boer, U., Eglins, J., Krause, D., Schnell, S., Schofl, C. and Knepel, W. (2007) Enhancement by lithium of cAMP-induced CRE/CREB-directed gene transcription conferred by TORC on the CREB basic leucine zipper domain. *Biochem J*, **408**, 69-77.

Bradford, M.M. (1976) A rapid and sensitive method for the quantitation of microgram quantities of protein utilizing the principle of protein-dye binding. *Anal Biochem*, **72**, 248-254.

Brocard, J.B., Rajdev, S. and Reynolds, I.J. (1993) Glutamate-induced increases in intracellular free Mg2+ in cultured cortical neurons. *Neuron*, **11**, 751-757.

Cade, J.F. (1949) Lithium salts in the treatment of psychotic excitement. *Med J Aust*, **2**, 349-352.

Canettieri, G., Coni, S., Della Guardia, M., Nocerino, V., Antonucci, L., Di Magno, L., Screaton, R., Screpanti, I., Giannini, G. and Gulino, A. (2009) The coactivator CRTC1 promotes cell proliferation and transformation via AP-1. *Proc Natl Acad Sci U S A*, **106**, 1445-1450.

Carlezon, W.A., Jr., Duman, R.S. and Nestler, E.J. (2005) The many faces of CREB. *Trends Neurosci*, **28**, 436-445.

Cerf, M.E. (2006) Transcription factors regulating beta-cell function. *Eur J Endocrinol*, **155**, 671-679.

Chawla, S., Hardingham, G.E., Quinn, D.R. and Bading, H. (1998) CBP: a signal-regulated transcriptional coactivator controlled by nuclear calcium and CaM kinase IV. *Science*, **281**, 1505-1509.

Chen, G., Hasanat, K.A., Bebchuk, J.M., Moore, G.J., Glitz, D. and Manji, H.K. (1999) Regulation of signal transduction pathways and gene expression by mood stabilizers and antidepressants. *Psychosom Med*, **61**, 599-617.

Chuang, D.M., Chen, R.W., Chalecka-Franaszek, E., Ren, M., Hashimoto, R., Senatorov, V., Kanai, H., Hough, C., Hiroi, T. and Leeds, P. (2002) Neuroprotective effects of lithium in cultured cells and animal models of diseases. *Bipolar Disord*, **4**, 129-136.

Coleman, D.E. and Sprang, S.R. (1998) Crystal structures of the G protein Gi alpha 1 complexed with GDP and Mg2+: a crystallographic titration experiment. *Biochemistry*, **37**, 14376-14385.

Colon-Cesario, W.I., Martinez-Montemayor, M.M., Morales, S., Felix, J., Cruz, J., Adorno, M., Pereira, L., Colon, N., Maldonado-Vlaar, C.S. and Pena de Ortiz, S. (2006) Knockdown of Nurr1 in the rat hippocampus: implications to spatial discrimination learning and memory. *Learn Mem*, **13**, 734-744.

Conkright, M.D., Canettieri, G., Screaton, R., Guzman, E., Miraglia, L., Hogenesch, J.B. and Montminy, M. (2003a) TORCs: transducers of regulated CREB activity. *Mol Cell*, **12**, 413-423.

Conkright, M.D., Guzman, E., Flechner, L., Su, A.I., Hogenesch, J.B. and Montminy, M. (2003b) Genome-wide analysis of CREB target genes reveals a core promoter requirement for cAMP responsiveness. *Mol Cell*, **11**, 1101-1108.

Craig, J.C., Schumacher, M.A., Mansoor, S.E., Farrens, D.L., Brennan, R.G. and Goodman, R.H. (2001) Consensus and variant cAMP-regulated enhancers have distinct CREB-binding properties. *J Biol Chem*, **276**, 11719-11728.

D'Elia, A.V., Puppin, C., Pellizzari, L., Pianta, A., Bregant, E., Lonigro, R., Tell, G., Fogolari, F., van Heyningen, V. and Damante, G. (2006) Molecular analysis of a human PAX6 homeobox mutant. *Eur J Hum Genet*, **14**, 744-751.

de Wet, J.R., Wood, K.V., DeLuca, M., Helinski, D.R. and Subramani, S. (1987) Firefly luciferase gene: structure and expression in mammalian cells. *Mol Cell Biol*, **7**, 725-737.

Dentin, R., Hedrick, S., Xie, J., Yates, J., 3rd and Montminy, M. (2008) Hepatic glucose sensing via the CREB coactivator CRTC2. *Science*, **319**, 1402-1405.

Dentin, R., Liu, Y., Koo, S.H., Hedrick, S., Vargas, T., Heredia, J., Yates, J., 3rd and Montminy, M. (2007) Insulin modulates gluconeogenesis by inhibition of the coactivator TORC2. *Nature*, **449**, 366-369.

Deutsch, P.J., Hoeffler, J.P., Jameson, J.L., Lin, J.C. and Habener, J.F. (1988) Structural determinants for transcriptional activation by cAMP-responsive DNA elements. *J Biol Chem*, **263**, 18466-18472.

Dwarki, V.J., Montminy, M. and Verma, I.M. (1990) Both the basic region and the 'leucine zipper' domain of the cyclic AMP response element binding (CREB) protein are essential for transcriptional activation. *Embo J*, **9**, 225-232.

Eckert, B., Schwaninger, M. and Knepel, W. (1996) Calcium-mobilizing insulin secretagogues stimulate transcription that is directed by the cyclic adenosine 3',5'-monophosphate/calcium response element in a pancreatic islet beta-cell line. *Endocrinology*, **137**, 225-233.

Einat, H., Yuan, P., Gould, T.D., Li, J., Du, J., Zhang, L., Manji, H.K. and Chen, G. (2003) The role of the extracellular signal-regulated kinase signaling pathway in mood modulation. *J Neurosci*, **23**, 7311-7316.

Escamilla, M.A. and Zavala, J.M. (2008) Genetics of bipolar disorder. *Dialogues Clin Neurosci*, **10**, 141-152.

Fang, H., Chartier, J., Sodja, C., Desbois, A., Ribecco-Lutkiewicz, M., Walker, P.R. and Sikorska, M. (2003) Transcriptional activation of the human brain-derived neurotrophic factor gene promoter III by dopamine signaling in NT2/N neurons. *J Biol Chem*, **278**, 26401-26409.

Felgner, P.L., Gadek, T.R., Holm, M., Roman, R., Chan, H.W., Wenz, M., Northrop, J.P., Ringold, G.M. and Danielsen, M. (1987) Lipofection: a highly efficient, lipid-mediated DNA-transfection procedure. *Proc Natl Acad Sci U S A*, **84**, 7413-7417.

Fiol, C.J., Williams, J.S., Chou, C.H., Wang, Q.M., Roach, P.J. and Andrisani, O.M. (1994) A secondary phosphorylation of CREB341 at Ser129 is required for the cAMP-mediated control of gene expression. A role for glycogen synthase kinase-3 in the control of gene expression. *J Biol Chem*, **269**, 32187-32193.

Flatman, P.W. (1991) Mechanisms of magnesium transport. *Annu Rev Physiol*, **53**, 259-271.

Foland, L.C., Altshuler, L.L., Sugar, C.A., Lee, A.D., Leow, A.D., Townsend, J., Narr, K.L., Asuncion, D.M., Toga, A.W. and Thompson, P.M. (2008) Increased volume of the amygdala and hippocampus in bipolar patients treated with lithium. *Neuroreport*, **19**, 221-224.

Frampton, J., Kouzarides, T., Doderlein, G., Graf, T. and Weston, K. (1993) Influence of the v-Myb transactivation domain on the oncoprotein's transformation specificity. *Embo J*, **12**, 1333-1341.

Frey, B.N., Andreazza, A.C., Nery, F.G., Martins, M.R., Quevedo, J., Soares, J.C. and Kapczinski, F. (2007) The role of hippocampus in the pathophysiology of bipolar disorder. *Behav Pharmacol*, **18**, 419-430.

Fukumoto, T., Morinobu, S., Okamoto, Y., Kagaya, A. and Yamawaki, S. (2001) Chronic lithium treatment increases the expression of brain-derived neurotrophic factor in the rat brain. *Psychopharmacology (Berl)*, **158**, 100-106.

Hartnett, K.A., Stout, A.K., Rajdev, S., Rosenberg, P.A., Reynolds, I.J. and Aizenman, E. (1997) NMDA receptor-mediated neurotoxicity: a paradoxical requirement for extracellular Mg2+ in Na+/Ca2+-free solutions in rat cortical neurons in vitro. *J Neurochem*, **68**, 1836-1845.

Hashimoto, K., Shimizu, E. and Iyo, M. (2004) Critical role of brain-derived neurotrophic factor in mood disorders. *Brain Res Brain Res Rev*, **45**, 104-114.

Hashimoto, M., Hsu, L.J., Rockenstein, E., Takenouchi, T., Mallory, M. and Masliah, E. (2002) alpha-Synuclein protects against oxidative stress via inactivation of the c-Jun N-terminal kinase stress-signaling pathway in neuronal cells. *J Biol Chem*, **277**, 11465-11472.

Herzig, S., Hedrick, S., Morantte, I., Koo, S.H., Galimi, F. and Montminy, M. (2003) CREB controls hepatic lipid metabolism through nuclear hormone receptor PPAR-gamma. *Nature*, **426**, 190-193.

Heukeshoven, J. and Dernick, R. (1988) Improved silver staining procedure for fast staining in PhastSystem Development Unit. I. Staining of sodium dodecyl sulfate gels. *Electrophoresis*, **9**, 28-32.

Heurteaux, C., Ripoll, C., Ouznadji, S., Ouznadji, H., Wissocq, J.C. and Thellier, M. (1991) Lithium transport in the mouse brain. *Brain Res*, **547**, 122-128.

Hu, S.C., Chrivia, J. and Ghosh, A. (1999) Regulation of CBP-mediated transcription by neuronal calcium signaling. *Neuron*, **22**, 799-808.

Impey, S., Fong, A.L., Wang, Y., Cardinaux, J.R., Fass, D.M., Obrietan, K., Wayman, G.A., Storm, D.R., Soderling, T.R. and Goodman, R.H. (2002) Phosphorylation of CBP mediates transcriptional activation by neural activity and CaM kinase IV. *Neuron*, **34**, 235-244.

Inaoka, Y., Yazawa, T., Uesaka, M., Mizutani, T., Yamada, K. and Miyamoto, K. (2008) Regulation of NGFI-B/Nur77 gene expression in the rat ovary and in leydig tumor cells MA-10. *Mol Reprod Dev*, **75**, 931-939.

Iourgenko, V., Zhang, W., Mickanin, C., Daly, I., Jiang, C., Hexham, J.M., Orth, A.P., Miraglia, L., Meltzer, J., Garza, D., Chirn, G.W., McWhinnie, E., Cohen, D., Skelton, J., Terry, R., Yu, Y., Bodian, D., Buxton, F.P., Zhu, J., Song, C. and Labow, M.A. (2003) Identification of a family of cAMP response element-binding protein coactivators by genome-scale functional analysis in mammalian cells. *Proc Natl Acad Sci U S A*, **100**, 12147-12152.

Johannessen, M., Delghandi, M.P. and Moens, U. (2004) What turns CREB on? *Cell Signal*, **16**, 1211-1227.

Jope, R.S. (1999) A bimodal model of the mechanism of action of lithium. *Mol Psychiatry*, **4**, 21-25.

Kato, T. (2007) Molecular genetics of bipolar disorder and depression. *Psychiatry Clin Neurosci*, **61**, 3-19.

Kato, T., Kakiuchi, C. and Iwamoto, K. (2007) Comprehensive gene expression analysis in bipolar disorder. *Can J Psychiatry*, **52**, 763-771.

Katoh, Y., Takemori, H., Lin, X.Z., Tamura, M., Muraoka, M., Satoh, T., Tsuchiya, Y., Min, L., Doi, J., Miyauchi, A., Witters, L.A., Nakamura, H. and Okamoto, M. (2006) Silencing the constitutive active transcription factor CREB by the LKB1-SIK signaling cascade. *Febs J*, **273**, 2730-2748.

Katoh, Y., Takemori, H., Min, L., Muraoka, M., Doi, J., Horike, N. and Okamoto, M. (2004) Salt-inducible kinase-1 represses cAMP response element-binding protein activity both in the nucleus and in the cytoplasm. *Eur J Biochem*, **271**, 4307-4319.

Kerppola, T.K. and Curran, T. (1993) Selective DNA bending by a variety of bZIP proteins. *Mol Cell Biol*, **13**, 5479-5489.

Knepel, W., Jepeal, L. and Habener, J.F. (1990) A pancreatic islet cell-specific enhancer-like element in the glucagon gene contains two domains binding distinct cellular proteins. *J Biol Chem*, **265**, 8725-8735.

Konig, H., Ponta, H., Rahmsdorf, U., Buscher, M., Schonthal, A., Rahmsdorf, H.J. and Herrlich, P. (1989) Autoregulation of fos: the dyad symmetry element as the major target of repression. *Embo J*, **8**, 2559-2566.

Koo, S.H., Flechner, L., Qi, L., Zhang, X., Screaton, R.A., Jeffries, S., Hedrick, S., Xu, W., Boussouar, F., Brindle, P., Takemori, H. and Montminy, M. (2005) The CREB coactivator TORC2 is a key regulator of fasting glucose metabolism. *Nature*, **437**, 1109-1111.

Kovacs, K.A., Steullet, P., Steinmann, M., Do, K.Q., Magistretti, P.J., Halfon, O. and Cardinaux, J.R. (2007) TORC1 is a calcium- and cAMP-sensitive coincidence detector involved in hippocampal long-term synaptic plasticity. *Proc Natl Acad Sci U S A*, **104**, 4700-4705.

Kovacs, K.J. (2008) Measurement of immediate-early gene activation- c-fos and beyond. *J Neuroendocrinol*, **20**, 665-672.

Kratzner, R., Frohlich, F., Lepler, K., Schroder, M., Roher, K., Dickel, C., Tzvetkov, M.V., Quentin, T., Oetjen, E. and Knepel, W. (2008) A peroxisome proliferator-activated receptor gamma-retinoid X receptor heterodimer physically interacts with the transcriptional activator PAX6 to inhibit glucagon gene transcription. *Mol Pharmacol*, **73**, 509-517.

Kruger, M., Schwaninger, M., Blume, R., Oetjen, E. and Knepel, W. (1997) Inhibition of CREB- and cAMP response element-mediated gene transcription by the immunosuppressive drugs cyclosporin A and FK506 in T cells. *Naunyn Schmiedebergs Arch Pharmacol*, **356**, 433-440.

Laemmli, U.K. (1970) Cleavage of structural proteins during the assembly of the head of bacteriophage T4. *Nature*, **227**, 680-685.

Lee, L.G., Connell, C.R. and Bloch, W. (1993) Allelic discrimination by nick-translation PCR with fluorogenic probes. *Nucleic Acids Res*, **21**, 3761-3766.

Lillie, J.W. and Green, M.R. (1989) Transcription activation by the adenovirus E1a protein. *Nature*, **338**, 39-44.

Liu, F. and Green, M.R. (1990) A specific member of the ATF transcription factor family can mediate transcription activation by the adenovirus E1a protein. *Cell*, **61**, 1217-1224.

Liu, Y., Dentin, R., Chen, D., Hedrick, S., Ravnskjaer, K., Schenk, S., Milne, J., Meyers, D.J., Cole, P., Yates, J., 3rd, Olefsky, J., Guarente, L. and Montminy, M. (2008) A fasting inducible switch modulates gluconeogenesis via activator/coactivator exchange. *Nature*, **456**, 269-273.

Lodish, H., Berk, A., Kaiser, C. A., and Matsudaira, P. (2004) Molecular Cell Biology, **5th Edition**, *Palgrave Macmillan, Houndmills, Basingstoke, Hampshire, England*.

Lonze, B.E. and Ginty, D.D. (2002) Function and regulation of CREB family transcription factors in the nervous system. *Neuron*, **35**, 605-623.

Maj, M. (2003) The effect of lithium in bipolar disorder: a review of recent research evidence. *Bipolar Disord*, **5**, 180-188.

Malenka, R.C. and Bear, M.F. (2004) LTP and LTD: an embarrassment of riches. *Neuron*, **44**, 5-21.

Manji, H.K., Moore, G.J. and Chen, G. (2000) Lithium up-regulates the cytoprotective protein Bcl-2 in the CNS in vivo: a role for neurotrophic and neuroprotective effects in manic depressive illness. *J Clin Psychiatry*, **61 Suppl 9**, 82-96.

Manji, H.K., Quiroz, J.A., Payne, J.L., Singh, J., Lopes, B.P., Viegas, J.S. and Zarate, C.A. (2003) The underlying neurobiology of bipolar disorder. *World Psychiatry*, **2**, 136-146.

Mantamadiotis, T., Lemberger, T., Bleckmann, S.C., Kern, H., Kretz, O., Martin Villalba, A., Tronche, F., Kellendonk, C., Gau, D., Kapfhammer, J., Otto, C., Schmid, W. and Schutz, G. (2002) Disruption of CREB function in brain leads to neurodegeneration. *Nat Genet*, **31**, 47-54.

Maxwell, M.A. and Muscat, G.E. (2006) The NR4A subgroup: immediate early response genes with pleiotropic physiological roles. *Nucl Recept Signal*, **4**, e002.

Mayr, B. and Montminy, M. (2001) Transcriptional regulation by the phosphorylation-dependent factor CREB. *Nat Rev Mol Cell Biol*, **2**, 599-609.

McClung, C.A. and Nestler, E.J. (2008) Neuroplasticity mediated by altered gene expression. *Neuropsychopharmacology*, **33**, 3-17.

Minadeo, N., Layden, B., Amari, L.V., Thomas, V., Radloff, K., Srinivasan, C., Hamm, H.E. and de Freitas, D.M. (2001) Effect of Li+ upon the Mg2+-dependent activation of recombinant Gialpha1. *Arch Biochem Biophys*, **388**, 7-12.

Miyamoto, E. (2006) Molecular mechanism of neuronal plasticity: induction and maintenance of long-term potentiation in the hippocampus. *J Pharmacol Sci*, **100**, 433-442.

Montezinho, L.P., C, B.D., Fonseca, C.P., Glinka, Y., Layden, B., Mota de Freitas, D., Geraldes, C.F. and Castro, M.M. (2004) Intracellular lithium and cyclic AMP levels are mutually regulated in neuronal cells. *J Neurochem*, **90**, 920-930.

Mota de Freitas, D., Castro, M.M. and Geraldes, C.F. (2006) Is competition between Li+ and Mg2+ the underlying theme in the proposed mechanisms for the pharmacological action of lithium salts in bipolar disorder? *Acc Chem Res*, **39**, 283-291.

Nair, A. and Vaidya, V.A. (2006) Cyclic AMP response element binding protein and brain-derived neurotrophic factor: molecules that modulate our mood? *J Biosci*, **31**, 423-434.

Nakajima, T., Uchida, C., Anderson, S.F., Parvin, J.D. and Montminy, M. (1997) Analysis of a cAMP-responsive activator reveals a two-component mechanism for transcriptional induction via signal-dependent factors. *Genes Dev*, **11**, 738-747.

Nekrep, N., Wang, J., Miyatsuka, T. and German, M.S. (2008) Signals from the neural crest regulate beta-cell mass in the pancreas. *Development*, **135**, 2151-2160.

Newman, M.E. and Belmaker, R.H. (1987) Effects of lithium in vitro and ex vivo on components of the adenylate cyclase system in membranes from the cerebral cortex of the rat. *Neuropharmacology*, **26**, 211-217.

Nordeen, S.K. (1988) Luciferase reporter gene vectors for analysis of promoters and enhancers. *Biotechniques*, **6**, 454-458.

Oetjen, E., Diedrich, T., Eggers, A., Eckert, B. and Knepel, W. (1994) Distinct properties of the cAMP-responsive element of the rat insulin I gene. *J Biol Chem*, **269**, 27036-27044.

Oetjen, E., Thoms, K.M., Laufer, Y., Pape, D., Blume, R., Li, P. and Knepel, W. (2005) The immunosuppressive drugs cyclosporin A and tacrolimus inhibit membrane depolarization-induced CREB transcriptional activity at the coactivator level. *Br J Pharmacol*, **144**, 982-993.

Omata, N., Murata, T., Takamatsu, S., Maruoka, N., Mitsuya, H., Yonekura, Y., Fujibayashi, Y. and Wada, Y. (2008) Neuroprotective effect of chronic lithium treatment against hypoxia in specific brain regions with upregulation of cAMP response element binding protein and brain-derived neurotrophic factor but not nerve growth factor: comparison with acute lithium treatment. *Bipolar Disord*, **10**, 360-368.

Pearse, A.G. and Polak, J.M. (1971) Neural crest origin of the endocrine polypeptide (APUD) cells of the gastrointestinal tract and pancreas. *Gut*, **12**, 783-788.

Pena de Ortiz, S., Maldonado-Vlaar, C.S. and Carrasquillo, Y. (2000) Hippocampal expression of the orphan nuclear receptor gene hzf-3/nurr1 during spatial discrimination learning. *Neurobiol Learn Mem*, **74**, 161-178.

Perlmann, T. and Wallen-Mackenzie, A. (2004) Nurr1, an orphan nuclear receptor with essential functions in developing dopamine cells. *Cell Tissue Res*, **318**, 45-52.

Peyton, M., Stellrecht, C.M., Naya, F.J., Huang, H.P., Samora, P.J. and Tsai, M.J. (1996) BETA3, a novel helix-loop-helix protein, can act as a negative regulator of BETA2 and MyoD-responsive genes. *Mol Cell Biol*, **16**, 626-633.

Phiel, C.J. and Klein, P.S. (2001) Molecular targets of lithium action. *Annu Rev Pharmacol Toxicol*, **41**, 789-813.

Plaumann, S., Blume, R., Borchers, S., Steinfelder, H.J., Knepel, W. and Oetjen, E. (2008) Activation of the dual-leucine-zipper-bearing kinase and induction of beta-cell apoptosis by the immunosuppressive drug cyclosporin A. *Mol Pharmacol*, **73**, 652-659.

Post, R.M. (2007) Role of BDNF in bipolar and unipolar disorder: clinical and theoretical implications. *J Psychiatr Res*, **41**, 979-990.

Pruunsild, P., Kazantseva, A., Aid, T., Palm, K. and Timmusk, T. (2007) Dissecting the human BDNF locus: bidirectional transcription, complex splicing, and multiple promoters. *Genomics*, **90**, 397-406.

Quiroz, J.A., Gould, T.D. and Manji, H.K. (2004) Molecular effects of lithium. *Mol Interv*, **4**, 259-272.

Rajkowska, G. (2002) Cell pathology in bipolar disorder. *Bipolar Disord*, **4**, 105-116.

Ramsby, M.L., Makowski, G.S. and Khairallah, E.A. (1994) Differential detergent fractionation of isolated hepatocytes: biochemical, immunochemical and two-dimensional gel electrophoresis characterization of cytoskeletal and noncytoskeletal compartments. *Electrophoresis*, **15**, 265-277.

Ravnskjaer, K., Kester, H., Liu, Y., Zhang, X., Lee, D., Yates, J.R., 3rd and Montminy, M. (2007) Cooperative interactions between CBP and TORC2 confer selectivity to CREB target gene expression. *Embo J*, **26**, 2880-2889.

Ryves, W.J., Dajani, R., Pearl, L. and Harwood, A.J. (2002) Glycogen synthase kinase-3 inhibition by lithium and beryllium suggests the presence of two magnesium binding sites. *Biochem Biophys Res Commun*, **290**, 967-972.

Ryves, W.J. and Harwood, A.J. (2001) Lithium inhibits glycogen synthase kinase-3 by competition for magnesium. *Biochem Biophys Res Commun*, **280**, 720-725.

Sadowski, I. and Ptashne, M. (1989) A vector for expressing GAL4(1-147) fusions in mammalian cells. *Nucleic Acids Res*, **17**, 7539.

Saiki, R.K., Gelfand, D.H., Stoffel, S., Scharf, S.J., Higuchi, R., Horn, G.T., Mullis, K.B. and Erlich, H.A. (1988) Primer-directed enzymatic amplification of DNA with a thermostable DNA polymerase. *Science*, **239**, 487-491.

Sambrook, J., Fritsch, E. F. and Maniatis, T. (1989) Molecular cloning. A laboratory manual, 2nd Edition, *Cold Spring Harbor Laboratory Press, Cold Spring Harbor, USA*.

Sanger, F., Nicklen, S. and Coulson, A.R. (1977) DNA sequencing with chain-terminating inhibitors. *Proc Natl Acad Sci U S A*, **74**, 5463-5467.

Santerre, R.F., Cook, R.A., Crisel, R.M., Sharp, J.D., Schmidt, R.J., Williams, D.C. and Wilson, C.P. (1981) Insulin synthesis in a clonal cell line of simian virus 40-transformed hamster pancreatic beta cells. *Proc Natl Acad Sci U S A*, **78**, 4339-4343.

Sassone-Corsi, P., Visvader, J., Ferland, L., Mellon, P.L. and Verma, I.M. (1988) Induction of proto-oncogene fos transcription through the adenylate cyclase pathway: characterization of a cAMP-responsive element. *Genes Dev*, **2**, 1529-1538.

Scherk, H., Reith, W. and Falkai, P. (2004) [Changes in brain structure in bipolar affective disorders]. *Nervenarzt*, **75**, 861-872.

Schreiber, E., Matthias, P., Muller, M.M. and Schaffner, W. (1989) Rapid detection of octamer binding proteins with 'mini-extracts', prepared from a small number of cells. *Nucleic Acids Res*, **17**, 6419.

Schumacher, M.A., Goodman, R.H. and Brennan, R.G. (2000) The structure of a CREB bZIP.somatostatin CRE complex reveals the basis for selective dimerization and divalent cation-enhanced DNA binding. *J Biol Chem*, **275**, 35242-35247.

Schwaninger, M., Blume, R., Kruger, M., Lux, G., Oetjen, E. and Knepel, W. (1995) Involvement of the Ca(2+)-dependent phosphatase calcineurin in gene transcription that is stimulated by cAMP through cAMP response elements. *J Biol Chem*, **270**, 8860-8866.

Schwaninger, M., Blume, R., Oetjen, E., Lux, G. and Knepel, W. (1993a) Inhibition of cAMP-responsive element-mediated gene transcription by cyclosporin A and FK506 after membrane depolarization. *J Biol Chem*, **268**, 23111-23115.

Schwaninger, M., Lux, G., Blume, R., Oetjen, E., Hidaka, H. and Knepel, W. (1993b) Membrane depolarization and calcium influx induce glucagon gene transcription in pancreatic islet cells through the cyclic AMP-responsive element. *J Biol Chem*, **268**, 5168-5177.

Screaton, R.A., Conkright, M.D., Katoh, Y., Best, J.L., Canettieri, G., Jeffries, S., Guzman, E., Niessen, S., Yates, J.R., 3rd, Takemori, H., Okamoto, M. and Montminy, M. (2004) The CREB coactivator TORC2 functions as a calcium- and cAMP-sensitive coincidence detector. *Cell*, **119**, 61-74.

Shaywitz, A.J. and Greenberg, M.E. (1999) CREB: a stimulus-induced transcription factor activated by a diverse array of extracellular signals. *Annu Rev Biochem*, **68**, 821-861.

Stemmer, W.P., Crameri, A., Ha, K.D., Brennan, T.M. and Heyneker, H.L. (1995) Single-step assembly of a gene and entire plasmid from large numbers of oligodeoxyribonucleotides. *Gene*, **164**, 49-53.

Stout, A.K., Li-Smerin, Y., Johnson, J.W. and Reynolds, I.J. (1996) Mechanisms of glutamate-stimulated Mg2+ influx and subsequent Mg2+ efflux in rat forebrain neurones in culture. *J Physiol*, **492 (Pt 3)**, 641-657.

Takemori, H. and Okamoto, M. (2008) Regulation of CREB-mediated gene expression by salt inducible kinase. *J Steroid Biochem Mol Biol*, **108**, 287-291.

Torii, T., Kawarai, T., Nakamura, S. and Kawakami, H. (1999) Organization of the human orphan nuclear receptor Nurr1 gene. *Gene*, **230**, 225-232.

Traut, T.W. (1994) Physiological concentrations of purines and pyrimidines. *Mol Cell Biochem*, **140**, 1-22.

Tsien, R.Y. (1998) The green fluorescent protein. *Annu Rev Biochem*, **67**, 509-544.

van Dam, H. and Castellazzi, M. (2001) Distinct roles of Jun : Fos and Jun : ATF dimers in oncogenesis. *Oncogene*, **20**, 2453-2464.

Vieta, E. and Sanchez-Moreno, J. (2008) Acute and long-term treatment of mania. *Dialogues Clin Neurosci*, **10**, 165-179.

Vinson, C., Myakishev, M., Acharya, A., Mir, A.A., Moll, J.R. and Bonovich, M. (2002) Classification of human B-ZIP proteins based on dimerization properties. *Mol Cell Biol*, **22**, 6321-6335.

Volpicelli, F., Caiazzo, M., Greco, D., Consales, C., Leone, L., Perrone-Capano, C., Colucci D'Amato, L. and di Porzio, U. (2007) Bdnf gene is a downstream target of Nurr1 transcription factor in rat midbrain neurons in vitro. *J Neurochem*, **102**, 441-453.

Webster, N., Jin, J.R., Green, S., Hollis, M. and Chambon, P. (1988) The yeast UASG is a transcriptional enhancer in human HeLa cells in the presence of the GAL4 trans-activator. *Cell*, **52**, 169-178.

Wells, J. and Farnham, P.J. (2002) Characterizing transcription factor binding sites using formaldehyde crosslinking and immunoprecipitation. *Methods*, **26**, 48-56.

Wine, Y., Cohen-Hadar, N., Freeman, A. and Frolow, F. (2007) Elucidation of the mechanism and end products of glutaraldehyde crosslinking reaction by X-ray structure analysis. *Biotechnol Bioeng*, **98**, 711-718.

Wu, H., Zhou, Y. and Xiong, Z.Q. (2007) Transducer of regulated CREB and late phase long-term synaptic potentiation. *Febs J*, **274**, 3218-3223.

Xing, G., Zhang, L., Russell, S. and Post, R. (2006) Reduction of dopamine-related transcription factors Nurr1 and NGFI-B in the prefrontal cortex in schizophrenia and bipolar disorders. *Schizophr Res*, **84**, 36-56.

Xu, W., Kasper, L.H., Lerach, S., Jeevan, T. and Brindle, P.K. (2007) Individual CREB-target genes dictate usage of distinct cAMP-responsive coactivation mechanisms. *Embo J*, **26**, 2890-2903.

Yucel, K., McKinnon, M.C., Taylor, V.H., Macdonald, K., Alda, M., Young, L.T. and MacQueen, G.M. (2007) Bilateral hippocampal volume increases after long-term lithium treatment in patients with bipolar disorder: a longitudinal MRI study. *Psychopharmacology (Berl)*, **195**, 357-367.

Yucel, K., Taylor, V.H., McKinnon, M.C., Macdonald, K., Alda, M., Young, L.T. and MacQueen, G.M. (2008) Bilateral hippocampal volume increase in patients with bipolar disorder and short-term lithium treatment. *Neuropsychopharmacology*, **33**, 361-367.

Zarate, C.A., Jr., Du, J., Quiroz, J., Gray, N.A., Denicoff, K.D., Singh, J., Charney, D.S. and Manji, H.K. (2003) Regulation of cellular plasticity cascades in the pathophysiology and treatment of mood disorders: role of the glutamatergic system. *Ann N Y Acad Sci*, **1003**, 273-291.

Zhou, Y., Wu, H., Li, S., Chen, Q., Cheng, X.W., Zheng, J., Takemori, H. and Xiong, Z.Q. (2006) Requirement of TORC1 for late-phase long-term potentiation in the hippocampus. *PLoS ONE*, **1**, e16.

Acknowledgements

I thank Professor W. Knepel for giving me the opportunity to work on this fascinating topic, allowing me to contribute to progress in understanding the molecular mechanism of the action of the mood stabilizer lithium. I am grateful for helpful advice and support in every phase of my work also under difficult circumstances.

I thank my thesis committee Professor G. Flügge, Professor R. Heinrich, and Professor K.-A. Nave for enriching discussion of my work and contribution to scientific progress during the development of my thesis.

Many thanks to Dr. Ulrike Böer and Dr. Elke Oetjen for the unreserved support of my work, for the scientific guidance and invaluable suggestions for the improvement of my thesis. I am grateful for your friendly care.

I am thankful to Mladen Tzvetkov for introducing me to the method of real-time PCR, for helpful discussions and practical advice.

I thank Doris Krause, Roland Blume, Corinna Dickel and Irmgard Chierny for their technical assistance and for encouraging guidance to learn new methods.

To all the actual and former members of the Department of Molecular Pharmacology, especially to Miranda, Helena, Catarina, Manuel, Phu, Svenja, Anne, Christine, Marcel, Andrei, Cordula and Sven: Many thanks to all of you for the nice working atmosphere, your helpfulness and support in many issues and for caring about my work.

Last, but not least, to Daniel Masuch, my family and all our friends: I thank you so much for believing in me and my work, for your understanding and support without any doubt.

Appendix A

Coding sequence of the TORC1 isoform cloned from HIT-T15 cells in the present study:

atggcgacttcgaacaatccgcggaaatttagcgagaagatcgcactgcacaaccagaagcaggcggaggagacggcggccttcga
ggaggtcatgaaggacctgagcctgacgcgggccgcgcggcttcaactgcagaagtcccagtacctgcagctgggccccagccgcgg
ccagtactatggcgggtccctgcccaacgtgaaccagattggaagtggcagcatggatttctccttccagacaccgtttcagtcctcaggcc
tggacacgagtcggaccacacgacaccatgggcttgtggacagagtgtatcgtgagaggggtagacttggctccccacaccgccggcc
cctgtcagttgacaagcacgggcgacaggctgacagctgcccctacggcaccgtgtacctctcgcctcctgcggacaccagctggagga
ggaccaactctgactctgccctgcaccagagcacaatgacacccacccaggcagaatccctctcaggcgggtcccaggacgcacatc
agaagagagtgttactgcttgccgtcccaggcatggaggagaccagatccgagacagacaagactctttctaagcagtcatgggactcc
aagaagacgggttccaggcccaagtcctgcgaggtccctggaatcaacatcttcccatctgcagagcaggagagcacggcagccctga
tccccgctccccacaacacaggggggctccctccctgacctcagcaacacgcatttccctccccactcccgacaccgctggaccccgag
gagcctgccttccctgctctcaccagctctggcagcaccggcagcctcgcacacctcggcgtcagcggcactgctcagggcatgaacac
ccccggctcttctccacagcaccggccagcagtcatcagcccctgtccctgagcacagaggccaggcggcagcaggcccagcaggt
gtcacccactctgtcccgttgtcacccatcactcaggccgtggccatggacgccctgtccctggaacaacagctgccctacgccttcttca
cccaggctggctcccagcagcctcccccacagcccagccaccacctccgcctccgcctgtgtccaacagcagccaccacccccac
aggtgtctgtgggcctcccccagggtggtccactgctgcccagtgccagcctgacgcggggggcccccagctgcctccgctctcaatcactgt
accgtccactctcccccagtctcctgcagagaaccctggccagccgccaatggggattgacaccacctcggcacaggctctgcagtacc
gcacgagcgcaggctcaccggccaaccagtctcccacgtctccagtctccaaccaaggcttctcccctggaagctccccacaacacac
atccaccctgggcagcgtgtttggggacgcattctatgagcagcagatgacagccaggcaggccaatgcgctgtcccgccagctggagc
agttcaacatgatggagaacgccatcagctccagcagcctgtacagcccaggctccactctgaactactcccaggctgccatgatgggc
ctgagcgggagccatggaggcctgcaggacccgcagcagctcagctacacaggccacagaggaatccccaacatcatcctcacggt
gacaggagagtcccccccagcctctctaaagaactgagcagctccctggcaggggtcagtgacgttagctttgactcggaccatcagtt
cccactggrtgaactga

I want morebooks!

Buy your books fast and straightforward online - at one of the world's fastest growing online book stores! Environmentally sound due to Print-on-Demand technologies.

Buy your books online at
www.get-morebooks.com

Kaufen Sie Ihre Bücher schnell und unkompliziert online – auf einer der am schnellsten wachsenden Buchhandelsplattformen weltweit!
Dank Print-On-Demand umwelt- und ressourcenschonend produziert.

Bücher schneller online kaufen
www.morebooks.de

OmniScriptum Marketing DEU GmbH
Heinrich-Böcking-Str. 6-8
D - 66121 Saarbrücken
Telefax: +49 681 93 81 567-9

info@omniscriptum.com
www.omniscriptum.com

Printed by Books on Demand GmbH, Norderstedt / Germany